NEXT WEEK

1. VOCABULARY TEST
2) DISCUSS CH I Questions
3) READ CH 2 pgs- 19-30

DISCUSS OHMS + WATTS LAWS

" CH 2 questions

NEXT WK- ~~READ CH 3~~

" " - quiz of on CH. 4

read CH 5

WINTER " CH 6 - TEST A/C BOOK - CH 7 up to motor protection ✓
(1ST HALF) devices

CH 7 - READ ⎫ TEST
CH-8 - READ ⎬

-SPRING-

✓CH 5 - 7133 - COOLING ELECTRIC - THURS ⎫ TEST
✓CH 9 +10- " " " " ⎬

✓CH 11 - THURS
✓CH -12 - THURS

FINAL - SCHMATIC - COOLING

Electricity For Refrigeration, Heating, and Air Conditioning

Russell E. Smith
Athens Technical School

Breton Publishers
North Scituate, Massachusetts

Library of Congress Cataloging in Publication Data

Library of Congress Cataloging in Publication Data

Smith, Russell E
  Electricity for refrigeration, heating, and air
conditioning.

  Includes index.
  1. Electric engineering. 2. Heating. 3. Air
conditioning. 4. Refrigeration and refrigerating
machinery. I. Title.
TK153.S57   621.3'02'4697   77-26751
ISBN 0-87872-147-9

Breton Publishers
A Division of Wadsworth, Inc.

©1978 by Wadsworth Publishing Company, Inc., Bel-
mont, California 94002. All rights reserved. No part of
this book may be reproduced, stored in a retrieval system,
or transcribed, in any form or by any means, electronic,
mechanical, photocopying, recording, or otherwise, without the prior written permission of the publisher, Breton
Publishers, 6 Bound Brook Court, North Scituate MA
02060, a division of Wadsworth, Inc.

Electricity for Refrigeration, Heating and Air Conditioning
was edited and prepared for composition by Carol Beal.
Interior design was provided by Amato Prudente and the
cover was designed by Oliver Kline.

L.C. Cat. Card No.: 77-26751

ISBN 0-87872-147-9

Printed in the United States of America

6   7   8   9 - 82

Contents

Preface   vii

1. Basic Electricity   1

1.1 Atomic Theory / 1   1.2 Positive and Negative
Charges / 3   1.3 Flow of Electrons / 4   1.4 Conductors and Insulators / 7   1.5 Electric Potential / 8
1.6 Current Flow / 10   1.7 Resistance / 11
1.8 Electric Power and Energy / 12   1.9 Ohm's
Law / 13   1.10 Calculating Electric Power / 16
Summary / 17   Questions / 18

2. Electric Circuits and Electric Meters   19

2.1 Basic Concepts of Electric Circuits / 20   2.2 Series
Circuits / 21   2.3 Parallel Circuits / 25   2.4 Series-
Parallel Circuits / 29   2.5 Electric Meters / 31
2.6 Ammeters / 34   2.7 Voltmeters / 37   2.8
Ohmmeters / 38   Summary / 41   Questions / 42

3. Components, Symbols, and Circuitry of Air-conditioning
Wiring Diagrams   44

3.1 Loads / 45   3.2 Contactors and Relays / 50
3.3 Magnetic Starters / 53   3.4 Switches / 53
3.5 Safety Devices / 58   3.6 Transformers / 60
3.7 Schematic Diagrams / 60   3.8 Reading Simple
Schematic Diagrams / 62   3.9 Reading Advanced
Schematic Diagrams / 68   3.10 Pictorial Diagrams / 74
3.11 Installation Diagrams / 76   Summary / 76
Questions / 79

4. **Alternating Current, Power Distribution, and Voltage Systems**   81

4.1 Basic Concepts of Alternating Current / 82   4.2 Power Distribution / 86   4.3 230 Volt-Single Phase-60 Hertz Systems / 87   4.4 Three-Phase Voltage Systems / 89   4.5 230 Volt-Three Phase-60 Hertz Systems / 90   4.6 208 Volt-Three Phase-60 Hertz Systems / 91   4.7 Higher-Voltage Systems / 92 Summary / 95   Questions / 96

5. **Installation of Heating, Cooling, and Refrigeration Systems**   98

5.1 Sizing Wire / 98   5.2 Disconnect Switches / 107 5.3 Fusible Load Centers / 110   5.4 Breaker Panels / 112   5.5 Distribution Centers / 116   Summary / 117 Questions / 118

6. **Basic Electric Motors**   119

6.1 Magnetism / 119   6.2 The Basic Electric Motor / 122   6.3 Types of Electric Motors / 125   6.4 Shaded-Pole Motors / 126   6.5 Split-Phase Motors / 129 6.6 Capacitors / 133   6.7 Permanent Split-Capacitor Motors / 135   6.8 Capacitor-Start Motors / 138 6.9 Capacitor-Start–Capacitor-Run Motors / 140   6.10 Three-Phase Motors / 141   6.11 Hermetic Compressor Motors / 144   Summary / 146   Questions / 147

7. **Components for Electric Motors**   149

7.1 Starting Relays for Single-Phase Motors / 150 7.2 Current or Amperage Relays / 150   7.3 Potential Relays / 152   7.4 Hot-Wire Relays / 154   7.5 Motor Bearings / 155   7.6 Motor Drives / 158   7.7 Magnetic Starters / 160   7.8 Push-button Stations / 163 Summary / 164   Questions / 165

8. **Contactors, Relays, and Overloads**   166

8.1 Contactors / 167   8.2 Relays / 173   8.3 Overloads / 177   Summary / 184   Questions / 185

9. Thermostats, Pressure Switches, and Other Electric Control Devices   **186**

9.1 Thermostats / 187   9.2 Staging Thermostats / 199 9.3 Pressure Switches / 202   9.4 Transformers / 207 9.5 Miscellaneous Electric Components / 210   Summary / 214   Questions / 215

10. Troubleshooting Electric Control Devices   **217**

10.1 Contactors and Relays / 218   10.2 Overloads / 220   10.3 Thermostats / 225   10.4 Pressure Switches / 227   10.5 Transformers / 229   10.6 Electric Motors / 230   Summary / 230   Questions / 231

11. Air-Conditioning Control Systems   **232**

11.1 Basic Condensing Units / 233   11.2 Packaged Units / 239   11.3 Field Wiring / 247   Summary / 254   Questions / 255

12. Control Systems: Circuitry and Troubleshooting   **257**

12.1 Basic Control Circuits / 258   12.2 Total Control System of Residential Units / 262   12.3 Advanced Control Systems / 263   12.4 Methods of Control on Advanced Systems / 267   12.5 Total Commercial and Industrial Control Systems / 270   12.6 Troubleshooting Control Systems / 272   Summary / 282   Questions / 283

Glossary   **287**

Index   **295**

Preface

Over the past several years as an instructor in the air-conditioning and refrigeration field, I have found it difficult to locate a text that gives the technician the necessary electrical background in terms that a person can readily grasp. Most books seem to be geared to the design engineer and are thus too mathematical or intended strictly for service personnel and thus too superficial. This book is written with a blend of theory and practice suitable for the vocational-technical student or the industry practitioner who wishes to upgrade his or her background.

The primary purpose of this text is to assemble those concepts and procedures that will enable the reader to work successfully in the industry. Electric principles, components, meters, schematics, and systems are discussed and applied to modern small- and large-scale installations. Troubleshooting and servicing are presented in practical terms in hopes of ensuring immediate productivity. This book does not replace manufacturers' service manuals but rather complements them.

The organization flows from the most basic units on the composition of electricity through increasingly larger components and ends with large-scale systems. Each chapter contains a summary and review questions.

As an expression of appreciation the author wishes to acknowledge the following: Athens Technical School, especially the students of the heating and air conditioning department, the staff, and the director, Mr. Robert G. Shelnutt.

Several people at Duxbury Press provided the motivation and technical expertise necessary to make this book happen. Ed Francis provided the initial inspiration and developmental assistance. Carol Beal did a masterful job of copy editing. Margaret Kearney provided able assistance in obtaining permissions and coordinating the art with the manuscript.

Finally, but not least, I would like to thank my wife for her patience and help.

1

Basic Electricity

INTRODUCTION

It would be extremely hard for anyone living today to be unaware of the importance of electricity to modern society. In any direction a person looks, electricity can be seen at work supplying the needs and luxuries of modern life. Electricity allows people to use light bulbs for light, motors to power rotating machinery, and electric heaters for heat. Electricity is one of the most important technological advances made in the history of the world.

Without electricity the heating, cooling, and refrigeration industry would never have advanced beyond its infancy. The industry relies almost totally on electricity for the operation of any heating, cooling, or refrigeration system. In any compression type of refrigeration system, there must be a means of rotating a compressor. In most cases the means is an electric motor. The industry also uses electricity for the control of all automatic systems.

Along with all the electric devices used in systems today come problems that are, in most cases, electrical and that must be corrected by field service personnel. Thus it is essential for all industry personnel to understand the basic principles of electricity so that they are able to perform their jobs in the industry.

We begin our study of electricity with a discussion of atomic structure.

1.1 ATOMIC THEORY

Matter is the substance of which a physical object is composed, whether it be a piece of iron, wood, or cloth, or whether it be a gas, a liquid, or a solid.

Matter is composed of fundamental substances called **elements**. There are 105 elements that have been found in the universe. Elements, in turn, are composed of atoms. An **atom** is the smallest particle of an element that can exist alone or in combination. All matter is made up of atoms or a combination of atoms. And all atoms are electrical in structure.

Suppose a piece of chalk is broken in half and one piece discarded. Then the remaining piece is broken in half and one piece discarded. If this procedure is kept up, eventually the piece of chalk will be broken into such a small piece that by breaking it once more there will no longer be a piece of chalk but only a molecule of chalk. A **molecule** is the smallest particle of a substance that has the properties of that substance. If a molecule of chalk is broken down into smaller segments, only individual atoms will exist, and they will no longer have the properties of chalk. The atom is the basic building block of all matter. The atom is the smallest particle that can combine with other atoms to form molecules.

Although the atom is a very small particle, it is also composed of several parts. The central part is called the **nucleus**. Another part, called **electrons**, orbits around the nucleus. The electron is a relatively small, negatively charged particle. The electrons orbit the nucleus in much the same way that the planets orbit the sun.

The nucleus, the center section of an atom, is composed of protons and neutrons. The **proton** is a heavy, positively charged particle. The proton has an electric charge that is opposite but equal to that of the electron. All atoms contain a like number of protons and electrons. The **neutron** is a neutral particle, which means that it is neither positively nor negatively charged. The neutrons tend to hold the protons together in the nucleus.

The simplest atom that exists is the hydrogen atom, which consists of one proton that is orbited by one electron, as shown in figure 1.1(a). All atoms are not as simple as the hydrogen atom. Other atoms have more particles. The difference in each different atom is the number of electrons, neutrons, and protons that the atom contains. The hydrogen atom has one proton and one electron. The oxygen atom has 8 protons, 8 neutrons, and 8 electrons, as shown in figure 1.1(b). The silver atom contains 47 protons, 61 neutrons, and 47 electrons. The more particles an atom has, the heavier the atom is. Since there are 105 elements, but millions of different types of substances, there must be some way of combining atoms and elements to form these substances.

When elements (and atoms) are combined they form a chemical union that results in a new substance, called a compound. For example, when two hydrogen atoms combine with one oxygen atom, the compound water is formed. The atomic structure of one molecule of water is shown in figure 1.1(c).

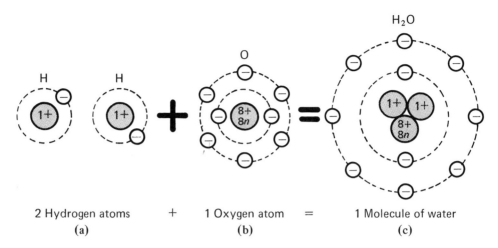

2 Hydrogen atoms   +   1 Oxygen atom   =   1 Molecule of water
(a)   (b)   (c)

FIGURE 1.1.   Atomic structure of a water molecule (one atom of oxygen and two atoms of hydrogen).

The chemical symbol for a compound denotes the atoms that make up
that compound. Refrigerant 12 is a substance commonly used in refrigeration systems. A refrigerant is a fluid that absorbs heat inside the conditioned
area and releases heat at a place outside the conditioned area. The chemical
symbol for one molecule of refrigerant 12 is CCl_2F_2. One molecule of the
refrigerant contains one atom of carbon, two atoms of fluorine, and two
atoms of chlorine. The chemical name for refrigerant 12 is dichlorodifluoro-
methane. All materials can be identified according to their chemical makeup,
that is, the atoms that form their molecules.

1.2 POSITIVE AND NEGATIVE CHARGES

An atom usually has an equal number of protons and electrons. When this
condition exists the atom is electrically neutral because the positively charged
protons exactly balance with the negatively charged electrons. However,
under certain conditions an atom can become unbalanced by losing or gaining
an electron. When an atom loses or gains an electron, it is no longer neutral.
It is either negatively or positively charged, depending on whether the electron is gained or lost. Thus in an atom a charge exists when the number of
protons and electrons is not equal.
    Under certain conditions some atoms can lose a few electrons for short
periods of time. Electrons that are in the outer orbits of some materials,
especially metals, can be easily knocked out of their orbits. Such electrons

3

are referred to as **free electrons**, and materials with free electrons are called **conductors**. When electrons are removed from the atom, the atom becomes positively charged because the negatively charged electrons have been removed, creating an unbalanced condition in the atom.

An atom can just as easily acquire additional electrons. When this occurs the atom becomes negatively charged.

Charges are thus created when there is an excess of electrons or protons in an atom. When one atom is charged and there is an unlike charge in another atom, electrons can flow between the two. This electron flow is called **electricity**.

An atom that has lost or gained an electron is considered unstable. A surplus of electrons in an atom creates a negative charge. A shortage of electrons creates a positive charge. Electric charges react to each other in different ways. Two negatively charged particles repel each other. Positively charged particles also repel each other. Two opposite charges attract each other. The **law of electric charges** states this: Like charges repel and unlike charges attract. Figure 1.2 shows an illustration of the law of electric charges.

All atoms tend to remain neutral because the outer orbits of electrons repel other electrons. However, many materials can be made to acquire a positive or negative charge by some mechanical means, such as friction. The familiar crackling when a hard rubber comb is run through hair on a dry winter day is an example of an electric charge generated by friction.

1.3   FLOW OF ELECTRONS

The flow of electrons can be accomplished by several different means: friction, which produces static electricity; chemical, which produces electricity in a battery; and magnetic (induction), which produces electricity in a generator. Other methods are also used, but the three we have mentioned here are the most common.

Static Electricity

The oldest known method of moving electrons is by **static electricity**. Static electricity produces a flow of electrons by permanently displacing an electron from an atom. The main characteristic of static electricity is that a prolonged or steady flow of current is not possible. As soon as the charges between the two substances are equalized (balanced), electron flow stops.

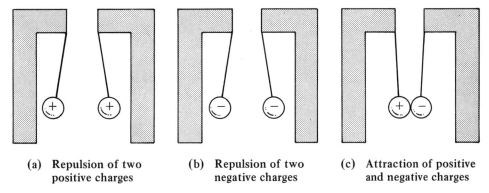

(a) Repulsion of two positive charges

(b) Repulsion of two negative charges

(c) Attraction of positive and negative charges

FIGURE 1.2.   Like charges repel and unlike charges attract each other.

Friction is usually the cause of static electricity. Sliding on a plastic seat cover in cold weather or rubbing silk cloth on a glass rod are two examples of static electricity produced by friction. Static electricity, no matter what the cause, is merely the permanent displacement or transfer of electrons. To obtain useful work from electricity, a constant and steady flow of electrons must be produced.

Electricity Through Chemical Means

Electricity can also be produced by the movement of electrons due to chemical means. A battery produces an electron flow by a chemical reaction that causes a transfer of electrons between two **electrodes**. One electrode collects electrons and one gives away electrons. The dry cell battery uses two electrodes made of two dissimilar metals inserted in a pastelike **electrolyte**. Electricity is produced when a chemical reaction occurs in the electrolyte between the electrodes, causing an electron flow. The construction of a dry cell battery is shown in figure 1.3.

The container of a dry cell battery, which is made of zinc, is the negative electrode (gives away electrons). The carbon rod in the center of the dry cell is the positive electrode (collects electrons). The space between the electrodes is filled with an electrolyte, usually manganese dioxide paste. The acid paste causes a chemical reaction between the carbon electrode and the zinc case. This reaction displaces the electrons, causing an electron flow. The top of the dry cell is sealed to prevent the electrolyte from drying and to allow the cell to be used in any position. The dry cell battery will eventually lose all its power because energy is being used and not being replaced.

The storage battery is different from a dry cell battery because it can be

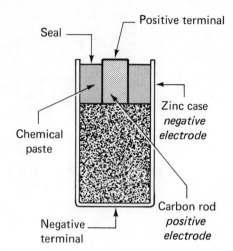

FIGURE 1.3.   Construction of a dry cell battery.

recharged. Thus it lasts somewhat longer than a dry cell battery. But it, too, will eventually lose all its energy.

The storage battery consists of a liquid electrolyte and negative and positive electrodes. The electrolyte is diluted sulfuric acid. The positive electrode is coated with lead dioxide and the negative electrode is sponge lead. The chemical reaction between the two electrodes and the electrolyte displaces electrons and creates voltage between the plates. The storage battery is recharged by reversing the current flow into the battery. The storage battery shown in figure 1.4 is commonly used in automobile electric systems.

Electricity Through Magnetism

The magnetic or induction method of producing electron flow uses a conductor to cut through a magnetic field, which causes a displacement of electrons. The alternator, generator, and transformer are the best examples of the magnetic method. The magnetic method is used to supply electricity to consumers.

The flow of electrons in a circuit produces magnetism, which is used to cause movement, or thermal energy, which in turn is used to cause heat. A magnetic field is created around a conductor—an apparatus for electrons to flow through—when there is a flow of electrons in the conductor. The flow of electrons through a conductor with a resistance will cause heat, such as in an electric heater.

The heating, cooling, and refrigeration industry uses magnetism to close

FIGURE 1.4. Common storage battery used in automobile electric systems.

relays and valves and to operate motors by using coils of wire to increase the strength of the magnetic field.

1.4 CONDUCTORS AND INSULATORS

The structure of an atom of an element is what makes it different from the atom of another element. The number of protons, neutrons, and electrons and the arrangement of the electrons in their orbits vary from element to element. In some elements the outer electrons rotating around the nucleus are easily removed from their orbits. As we stated earlier, elements that have atoms with this characteristic and called **conductors**. A conductor can transmit electricity or electrons.

Most metals are conductors, but all metals do not conduct electricity equally well. The best conductors are silver, copper, and aluminum. The high cost of silver keeps it from being used widely. Its use is largely limited to contacts in certain electrical switching devices such as contactors and relays. Copper, almost as good a conductor as silver, is usually used because it is less expensive.

Materials that do not easily give up or take on electrons are called **insulators**. An insulator retards the flow of electrons. Glass, rubber, and asbestos are examples of insulators. How well an insulator prevents electron flow depends on the strength of that flow. If the flow of electrons is strong enough, the insulator will break down, causing electrons to flow through it.

There is no perfect insulator. All insulators will break down under certain conditions if the electron flow is high enough. Increasing the thickness of the insulation helps overcome this problem.

7

Conductors and insulators are important parts of electric circuits and electric systems. They are widely used in all electric components in the industry.

1.5 ELECTRIC POTENTIAL

In a water system water can flow as long as pressure is applied to one end of a pipe and the other end of the pipe is open. The greater the pressure in a water system, the greater the quantity of water that will flow. Similarly, in an electrical system electrons will flow as long as electric pressure is applied to the system. **Voltage, potential difference**, and **electromotive force** are all terms used to describe electric pressure.

Recall that the law of electric charges states that unlike charges attract. Consequently there is a pull, or *force,* of attraction between two dissimilarly charged objects. We call this pull of attraction a **field of force**.

Another way of looking at this is to picture excess electrons (the negative charge) as straining to reach the point where there are not enough electrons (the positive charge). If the two charges are connected by a conductor, the excess electrons will flow to the point where there are not enough electrons. But if the two charges are separated by an insulator, which prevents the flow of electrons, the excess electrons cannot move. Hence an excess of electrons will pile up at one end of the insulator, with a corresponding lack, or deficiency, of electrons at the other end.

As long as the electrons cannot flow, the field of force between the two dissimilarly charged ends of the insulator increases. The resulting strain between the two ends is called the **electric pressure**. This pressure can become quite great. After a certain limit is reached, the insulator can no longer hold back the excess electrons, as we discussed in the previous section. Hence the electrons will rush across the insulator to the other end.

Electric pressure that causes electrons to flow is called voltage. Voltage is the difference in electric potential (or electric charge) between two points. The **volt** (V) is the amount of pressure required to force one ampere (A, the unit of measurement for current flow) through a resistance of one ohm (Ω, the unit of measurement for resistance; Ω is the Greek letter omega). In the industry, voltage is almost always measured in the range of the common volt. In other areas the volt may be measured on a smaller scale of millivolt (mV), or one-thousandth of a volt. For larger measurements of the volt, the kilovolt (kV), equal to 1000 volts, is used.

$$1 \text{ millivolt} = 0.001 \text{ volt}$$

$$1 \text{ kilovolt} = 1000 \text{ volts}$$

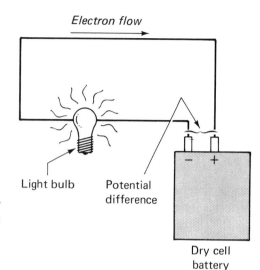

FIGURE 1.5. A dry cell battery supplying electric potential (voltage) to an
electric circuit.

To maintain electric pressure we must have some means to move electrons in the same manner that water pressure moves water. In an electric
circuit this can be maintained by a battery, as shown in figure 1.5, or by a
generator or alternator, as shown in figure 1.6. The battery forces electrons
to flow to the positive electrode and causes electric pressure. A generator
causes electric pressure by transferring electrons from one place to another.

Electromotive force can be produced in several different ways. The easiest method to understand is the simple dry cell battery discussed in section
1.3. The most popular method of producing an electromotive force is by
using an alternating current generator. The alternating current generator is
supplied with power from another source. Then a wire loop is rotated through
the magnetic field created by the voltage being applied, and an electromotive
force is produced through the wire loop. We will discuss these ideas in more
detail in succeeding sections.

FIGURE 1.6. A generator supplying
electric potential (voltage) to an electric
circuit.

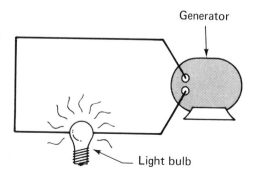

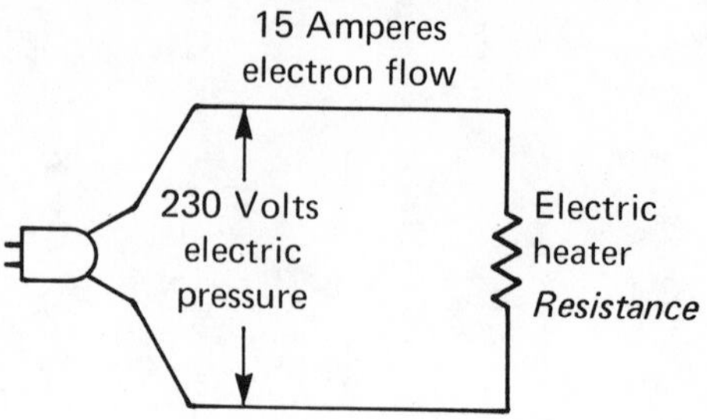

15 Amperes
electron flow

230 Volts
electric
pressure

Electric
heater
Resistance

FIGURE 1.7. An electric system; electric potential forces electrons through a wire (a conductor).

1.6 CURRENT FLOW

Electrons flowing in an electric circuit are called **current**. Current flow can be obtained in an electric circuit by a bolt of lightning, by static electricity, or by electron flow from a generator. Figure 1.7 shows an electric system with electric pressure; the quantity of electrons flowing is also given.

There are two types of electric current: **direct current** and **alternating current**. Direct current flows in one direction. It is the type of current produced by dry cell batteries. Its use is rare in the industry, except for some of the solid-state modules now being used.

Alternating current flows back and forth. It is the type of current available in most homes supplied by an electric utility company. It is the most commonly used source of electric potential in the heating, cooling, and refrigeration industry. Alternating current will be discussed in more detail in chapter 4.

The current in an electric circuit is measured in **amperes** (A). An ampere is the amount of current required to flow through a resistance of one ohm with a pressure of one volt. An ampere is measured with an **ammeter**. In the industry the ampere is used almost exclusively. If a smaller unit of ampere measurement is required, the milliampere (mA), which is one-thousandth of an ampere, can be used. For larger measurements of amperes, the kiloampere (kA) can be used. One kiloampere equals 1000 amperes.

1 milliampere = 0.001 ampere

1 kiloampere = 1000 amperes

The current that an electric device consumes can be used as a guide to the correct operation of the equipment by installation and service personnel. The electric motor is the largest current-consuming device in most heating, cooling, and refrigeration systems. The larger the electric device (load), the larger the current flow. Any electric device that uses electricity requires a certain current when operating properly.

1.7 RESISTANCE

The **resistance** to the flow of electrons in an electric circuit is measured in ohms. Figure 1.8 shows two electric systems with different resistances. One ohm is the amount of resistance that will allow one ampere to flow with a pressure of one volt. The industry uses simple ohms for resistance in most cases because the scale is broad enough for most applications. In some special cases the microhm, ($\mu\Omega$), which is one-millionth of an ohm, is used for extremely small resistance readings. Larger resistance readings are read in megohms (MΩ); one megohm is equal to one million ohms.

$$1 \text{ microhm} = 0.000001 \text{ ohm}$$

$$1 \text{ megohm} = 1,000,000 \text{ ohms}$$

All electric devices will have a certain resistance. That resistance depends on the size and purpose of the device. As service personnel you will have to become familiar with this value in components. If the resistance deviates far from the specified value or the estimated value, the device can be considered faulty.

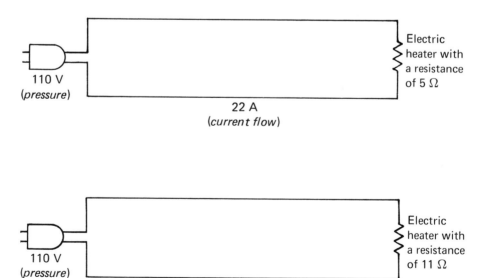

FIGURE 1.8. *Two electric systems with different resistances.*

1.8 ELECTRIC POWER AND ENERGY

When electrons move from the negative to the positive end of a conductor, work is done. **Electric power** is the rate at which the electrons do work. That is, electric power is the rate at which electricity is being used. The power of an electric circuit is measured in **watts** (W). A watt of electricity is one ampere flowing with a pressure of one volt. The electric power of a circuit is the voltage times the amperage.

$$\text{electric power} = \text{voltage} \times \text{amperage}$$

In an alternating current circuit, the voltage and current are not in phase. To obtain the correct power consumed by the circuit, the product of the voltage times the amperage must also be multiplied by a **power factor**. The power factor is the power the utility company gets paid for divided by the power the utility company produces. It is expressed as a percentage. In a direct current circuit, the product of the voltage times the amperage gives the power of the circuit; the power factor is not needed.

The industry uses the units of watts for devices that consume a small amount of power. Examples of such devices are small electric motors and small resistance heaters. Other units used are the horsepower (hp) and the British thermal unit (Btu). One horsepower is equal to 746 watts. One watt is equal to 3.41 Btu per hour.

$$1 \text{ horsepower} = 746 \text{ watts}$$
$$1 \text{ watt} = 3.41 \text{ Btu/hour}$$

These conversion figures are often used in the industry to calculate the Btu rating of an electric heater if the watt rating is known. The horsepower conversion is used to calculate the horsepower of a motor if only the watts are known.

The rate at which electric power is being used at a specific time is called **electric energy**. Electric energy is measured in watthours (Wh). For example, a one-horsepower motor uses 746 watts in one hour. It therefore uses 746 watthours.

The wattage rating or consumption of any electric device only denotes how much power the device is using. However, time must be considered when calculating electric energy, that is, the power being consumed over a definite period of time. Watthours give the number of watts used for a specific period of time.

The units of kilowatts (kW) are usually used to determine the amount of electricity consumed. Thus an electric utility calculates the power bill of its customers using **kilowatthours** (kWh) because the watthour readings would be extremely large. The kilowatthour reading is relatively small. One thousand watts used for one hour equals one kilowatthour. All electric meters used to measure the consumption of electricity record consumption in kilowatthours.

Heating and air-conditioning personnel are often required to make a calculation for the output of an electric heater in Btu's rather than in watts. The industry rates electric heating equipment in watts or kilowatts, which does not give consumers figures that they can understand. Most consumers are familiar with the term Btu and know basically what it means in terms of heat output. Therefore, the consumer often will request the Btu output rather than the wattage of a heating appliance. The Btu output can be easily calculated by multiplying the number of watts by the conversion factor of 3.41 Btu/h. Figure 1.9 shows an electric data plate for an electric heater. The wattage on the data plate is 4800 watts. Therefore the Btu output is

$$4,800 \times 3.41 \text{ Btu/h} = 16,368 \text{ Btu/h/stage}$$

Another term that is rapidly gaining popularity due to the energy crisis is the **energy efficiency rate** (EER). EER is the energy efficiency rating of an air-conditioning unit, measured in Btu per watt. For example, if an air-conditioning unit has an EER of 7.9, it will produce 7.9 Btu of cooling per watt of power consumed by the equipment. Almost all air-conditioning manufacturers are using EER ratings for their equipment.

1.9 OHM'S LAW

The relationship among the current, electromotive force, and resistance in an electric circuit is known as Ohm's law. In the nineteenth century George Ohm developed the mathematical comparisons of the major factors in an electric circuit. Stated in simple terms, Ohm's law says that the greater the voltage, the greater the current, and the greater the resistance, the lesser the current. Ohm's law is represented mathematically as "current is equal to the electromotive force divided by the resistance." The following equation expresses Ohm's law:

$$I = \frac{E}{R}$$

In the equation I represents the current in amperes, E represents the electromotive force in volts, and R represents the resistance in ohms. Ohm's law can also be expressed by the following two formulas:

$$E = IR \quad R = \frac{E}{I}$$

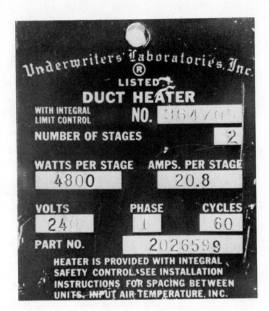

FIGURE 1.9. *Data plate of an electric heater.*

In any of the three formulas, when two elements of an electric circuit are known, the unknown factor can be calculated. Ohm's law does not apply to alternating current circuits because coils of wire produce different effects on alternating current. Alternating current will be discussed in a later chapter. However, the general ideas of Ohm's law apply to alternating current circuits.

The following examples show the relationships of the voltage, current, and resistance in an electric circuit.

Example 1   What is the current in the circuit shown in figure 1.10?

Step 1: $I = \dfrac{E}{R}$

Step 2: $I = \dfrac{120}{10}$

Step 3: $I = 12$ A

FIGURE 1.10. *Simple electric circuit for example 1.*

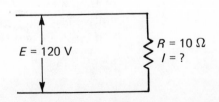

$E = 120$ V

$R = 10\ \Omega$
$I = ?$

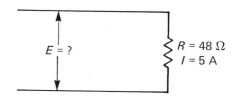

FIGURE 1.11. Simple electric circuit for example 3.

Example 2   What is the resistance of a 100-watt light bulb if the voltage is 120 volts and the current is .83 amperes?

Step 1: $R = \dfrac{E}{I}$

Step 2: $R = \dfrac{120}{.83}$

Step 3: $R = 145\ \Omega$

Example 3   What is the voltage supplied to the circuit in figure 1.11?

Step 1: $E = IR$

Step 2: $E = 5 \times 48$

Step 3: $E = 240\ \text{V}$

FIGURE 1.12. Using Ohm's law.

If calculating for E, cover E and use the formula

$$E = I \times R$$

If calculating for I, cover I and use the formula

$$I = \frac{E}{R}$$

If calculating for R, cover R and use the formula

$$R = \frac{E}{I}$$

Ohm's law allows the calculation of the missing factor if the other two factors are known or can be measured. Figure 1.12 shows a simple method for remembering Ohm's law. If one of the factors in the circle is covered, the letters remaining in the circle give the correct formula for calculating the covered factor.

1.10 CALCULATING ELECTRIC POWER

Electric power can be calculated by using the formula $P = IE$. Two other formulas can be used to calculate the electric power of an electric circuit by substituting in the following equations:

$$P = \frac{E^2}{R} \quad P = I^2 R$$

The letter designations in these formulas are the same as in Ohm's law, with P representing power in watts.

The following three examples show the electric power calculations of three electric circuits.

Example 4   What is the power consumption of an electric circuit using 15 amperes and 120 volts?

Step 1: $P = IE$

Step 2: $P = 15 \times 120$

Step 3: $P = 1800 \text{ W}$

Example 5   What is the current of an electric heater rated at 5000 watts on 230 volts?

Step 1: $I = \dfrac{P}{E}$

Step 2: $I = \dfrac{5000}{230}$

Step 3: $I = 21.7 \text{ A}$

Example 6   What is the power of an electric circuit with 5 amperes current and 10 ohms resistance?

Step 1: $P = I^2 R$

Step 2: $P = 5^2 \times 10$

Step 3: $P = 25 \times 10$

Step 4: $P = 250 \text{ W}$

SUMMARY

Everything—solids, liquids, and gases—is composed of matter. Matter can be broken down into molecules (the smallest particle of physical objects) and atoms (the smallest particle of an element that can exist alone or in combination). An atom is composed of a nucleus (the central part) and electrons (negatively charged) that orbit around the nucleus, much like the planets orbit the sun. The nucleus is composed of protons (positively charged) and neutrons (no charge). The number of protons is usually equal to the number of electrons, making the atom electrically neutral. When an atom loses electrons, it becomes positively charged. When it gains electrons, it becomes negatively charged. The law of charges states that like charges repel and unlike charges attract. Materials can be made to acquire positive or negative charges.

Electrons can be made to flow by use of friction, chemicals, and magnetism. A conductor is a material capable of transmitting electrons or electricity. Most metals are conductors. An insulator is a material that resists or prevents electron flow.

There are four important factors in any electric circuit: electromotive force, current, resistance, and power. The electromotive force of an electric circuit is the actual pressure in the circuit, much like water pressure in a water system. The electromotive force in an electric circuit is measured in volts. The voltage (pressure) must be sufficient to overcome the resistance of the circuit. Alternating current is used almost exclusively in the industry to supply electric power to equipment.

The number of electrons flowing in an electric circuit is called the current flow. The current flow of an electric circuit is measured in amperes. An ampere is the amount of current that will flow through a resistance of one ohm with a pressure of one volt.

The resistance of an electric circuit is measured in ohms. All electric loads have some resistance. Electric power is the rate at which electric energy is being used in an electric circuit. Electric power is measured in watts and kilowatts. The electric utilities use kilowatthours in most cases to charge their customers for the electric energy they have consumed. The kilowatthour is a measure of electric energy and takes into consideration the amount of time and the power consumption. One thousand watts used for a period of one hour equals one kilowatthour. Voltage, amperage, resistance, and wattage often use the prefixes kilo- or milli- to represent larger or smaller quantities of these factors and to avoid the use of extremely large or small numbers.

Ohm's law gives the relationship among the current, electromotive force, and resistance in an electric circuit. Ohm's law states the relationship mathematically. When any two factors in an electric circuit are known or can be measured, the formulas for Ohm's law can be used to find the third factor. Electric power can be calculated by using the formula $P = IE$.

QUESTIONS

1. What is an atom? What are the parts in an atom?

2. What is static electricity?

3. Name three ways electricity can be produced.

4. What parts do protons and electrons play in the production of electricity? *CHARGE* *create a*

5. The simplest atom that exists is _____.

6. What are the three most important factors in an electric circuit? *SOURCE CONDUCTOR LOAD*

7. What is electromotive force, and how is it measured?

8. What is current, and how is it measured?

9. What is resistance, and how is it measured?

10. What is electric power, and how is it measured?

11. Where do electrons exist in an atom, and what is their charge?

12. True or false. All atoms tend to lose electrons.

13. State the law of charges.

14. What is a proton? Where does it normally exist in an atom, and what is its charge?

15. Describe briefly the way a dry cell battery operates.

16. What is a conductor?

17. What is an insulator?

18. Why do metals make the best conductors? *LOSE ELECTRONS EASILY*

19. In a battery the transfer of electrons occurs between two _____.

20. How do electric utilities charge consumers for electricity?

21. What is the source of electric pressure in an electric circuit?

22. What is the meaning of EER when used in conjunction with an air-conditioning unit?

23. State Ohm's law.

24. Why does Ohm's law not apply to alternating current?

25. What is the ampere draw of a 5000-watt electric heater used on 110 volts?

26. What is the resistance of the heating element of an electric iron if the ampere draw is 8 amperes when 115 volts are applied?

27. What is the voltage applied to a small electric heater if the heater is pulling 12 amperes and has a resistance of 10 ohms?

28. What is the Btu/h output of an electric heater rated at 15 kW?

29. What is the kilowatt output of an electric heater that has an ampere draw of 50 A and a voltage source of (a) 208 V? (b) 230 V?

2

Electric Circuits and Electric Meters

INTRODUCTION

The circuitry of the electric system in today's heating, cooling, and refrigeration systems is very important to the personnel who install or work on the electric systems. Any electric system used in the industry today is composed of various types of circuits. Each type is designed to do a specific task within the system. We will look at some of the more commonly used types of circuits in this chapter.

The two most important kinds of circuits are parallel circuits and series circuits. A **parallel circuit** is an electric circuit that has more than one path through which electricity may flow. A parallel circuit is designed to supply more than one load in the system.

A **series circuit** is an electric circuit that has only one path through which electricity may flow. It is usually used for devices that are connected in the circuit for safety or control.

You must understand the circuitry in air-conditioning, heating, and refrigeration control and power systems to do an effective job of installing and servicing the equipment.

Electric meters also play an important part in the industry. All air-conditioning systems in common use today have some form of electricity that is used for control or operation. Hence industry personnel must become familiar with all types of electric meters to do their assigned tasks. The installation mechanic will have to be able to read and use electric meters in order to check the electrical characteristics of a newly installed air-conditioning or heating system. No installation is complete without complete electrical checks of the system. The service mechanic must know how to use electric meters because of the many electrical problems that occur in the field.

We begin our study with a discussion of the basics of electric circuits.

2.1   BASIC CONCEPTS OF ELECTRIC CIRCUITS

An **electric circuit** is the complete path of an electric current, along with any necessary elements, such as a power source and a load. When the circuit is complete so that the current can flow, it is termed **closed** or made (see figure 2.1). When the path of current flow is interrupted, the circuit is termed **open** or broken (see figure 2.2). The opening and closing of electric circuits control the operation of loads in the circuit.

All electric circuits must have a complete path for electrons to flow through, a source of electrons, and some electric device (load) that requires electric energy for its operation. Figure 2.3 shows a complete circuit with the basic components labeled. Electric circuits supply power and control loads through the use of switches.

An alternating current power supply is the most common source of the electric energy needed in the electric circuits of a heating, cooling, or refrigeration system. A direct current power supply, such as the dry cell battery, is often the source of electron flow in electric meters. Other than for meters, though, direct current is rarely used as a source of electric energy. Two special applications of direct current used today are electronic air cleaners and solid-state modules used for some types of special control, such as a defrost control or overcurrent protection.

The purpose of most electric circuits is supplying energy to a machine that does work. The most common device supplied with energy is an electric motor. Motors are used to rotate fans, compressors, pumps, and other me-

FIGURE 2.1.   A closed circuit.

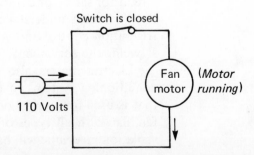

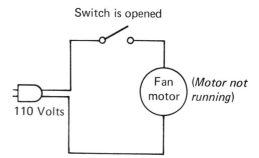

FIGURE 2.2. *An open circuit.*

chanical devices that require a rotating motion. Automatic switches also require a power source to open and close them so they can start other electric devices. Electric circuits also supply energy to transformers, lights, and timers.

There are several possible path arrangements that electrons may follow. The path arrangement is determined by the use or purpose of the circuit. The three types of circuits are series, parallel, and series-parallel, The series circuit allows only one path of electron flow. The parallel circuit has more than one path. The series-parallel circuit is a combination of the series and parallel circuits. In the following sections we will look at each of these arrangements.

2.2 SERIES CIRCUITS

The simplest and easiest electric circuit to understand is the series circuit. The series circuit allows only one path of current flow through the circuit. In other words, the path of a series circuit must pass through each and every load of the entire circuit. That is, all loads are connected end to end within

FIGURE 2.3. *Basic electric circuit with circuit components labeled.*

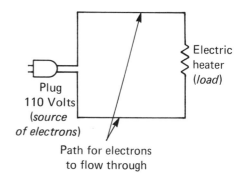

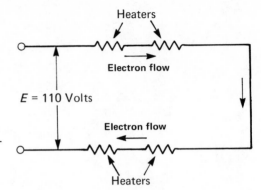

FIGURE 2.4. *Series circuit with four re-*
sistance heaters.

the series circuit. Figure 2.4 shows a series circuit with four resistance heaters.

Applications

Series circuits are incorporated in almost all control circuits used in heating, cooling, and refrigeration equipment. A **control circuit** is an electric circuit that controls some major load in the system. If all control components are connected in the circuit in series, the opening of any switch or component will open the circuit and stop the electric load, as shown in the circuit in figure 2.5. (The symbols L_1 and L_2 in figure 2.5 and in other figures represent the source of the voltage, the power supply. We will discuss circuit notation in more detail in chapter 3.)

Series circuits are used in the electric circuitry of heating, cooling, and refrigeration equipment to operate the equipment and maintain a desired temperature. Any electric switch or control that is placed in series with a

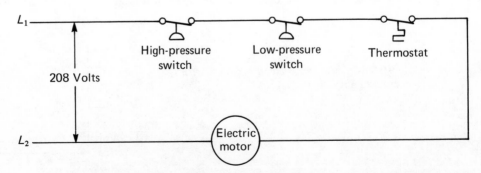

FIGURE 2.5. *Series circuit with three switches for controlling an elec-*
tric motor.

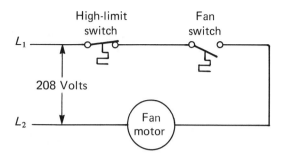

FIGURE 2.6. Control (series) circuit of a fan motor in a gas furnace; both temperature controls must be closed for the fan to operate.

load will operate that particular load. Figure 2.6 shows one circuit of a control system. The controls are connected in series with the device that is being controlled, in this case an electric motor.

The series circuit also contains any safety devices that are needed to maintain safe operation of the equipment components. Figure 2.7 shows a series circuit that is basically made up of safety devices designed to stop the equipment if an unsafe operating condition occurs. If any of the safety controls opens, the circuit will open and stop the equipment.

Calculations for Current, Resistance, and Voltage

The current draw in a series circuit is the same throughout the entire circuit because there is only one path for the current to follow. The current in a series circuit is shown by the following equation:

$$I_t = I_1 = I_2 = I_3 = I_4 = \cdots$$

(The centered dots $\cdots$ indicate that the equation continues in the same manner until all the elements of that particular circuit have been accounted for.)

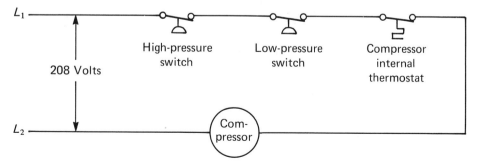

FIGURE 2.7. Control (series) circuit showing safety devices used on modern air-conditioning systems.

The total resistance R_t in a series circuit is the sum of all the resistances in the circuit. The resistance of a series circuit is shown by the following equation:

$$R_t = R_1 + R_2 + R_3 + R_4 + \cdots$$

The voltage in a series circuit is completely used by all the loads in the circuits. The loads of the series circuit must share the voltage that is being delivered to the circuit. Thus the voltage being delivered to the circuit will be split by the loads in the circuit.

The voltage of a series circuit changes through each load. This change is called the **voltage drop**. The voltage drop is the amount of voltage lost through any load or conductor. The voltage drop of any part of a series circuit is proportional to the resistance in that part of the circuit. The sum of the voltage drops of a series circuit is equal to the voltage being applied to the circuit. This is shown by the following equation:

$$E_t = E_1 + E_2 + E_3 + E_4 + \cdots$$

Ohm's law can be used for the calculations on any part of a series circuit or on the total circuit. Figure 2.8 shows a series circuit with four resistance heaters of different ohm ratings. The calculations for the total resistance, the amperage, and the voltage drop across each heater will be calculated using the circuit shown in figure 2.8.

The total resistance can be calculated by adding the ohm rating of each heater.

Step 1:   Use the formula $R_t = R_1 + R_2 + R_3 + R_4$.

Step 2:   Substitute the values given in the figure into the formula:

$$R_t = 4 \text{ ohms} + 10 \text{ ohms} + 12 \text{ ohms} + 14 \text{ ohms}$$

Step 3:   Solve the formula: $R_t = 40$ ohms

We use Ohm's law to calculate the amperage draw (current) of the circuit.

Step 1:   Use the formula

$$I = \frac{E}{R} \quad = \frac{120}{40} = 3 \text{ A.}$$

FIGURE 2.8. Series circuit containing four resistance heaters with different ohmic values.

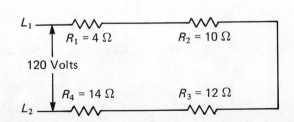

Step 2: Substitute the given values into the formula:

$$I = \frac{120}{40}$$

Step 3: Solve the formula: $I = 3$ amperes

Now we use Ohm's law to calculate the voltage drop across each heater.

Step 1: Use the formula $E = IR$ for each resistance.
Step 2: Substitute the given values into the formula: $E_{d1} = 3 \times 4$
   (The symbol E_{d1} means the voltage drop across resistance 1.)
Step 3: Solve the formula: $E_{d1} = 12$ volts
Step 4: Solve for each resistance using the same procedures that were used in steps 2
   and 3.

$E_{d1} = 3 \times 4 = 12$

$$E_{d2} = 3 \times 10 = 30 \text{ volts}$$
$$E_{d3} = 3 \times 12 = 36 \text{ volts}$$
$$E_{d4} = 3 \times 14 = 42 \text{ volts}$$

$120\,\sqrt{}$

Note that the total voltage E_t supplied to the circuit is equal to the sum of the voltage
drops.

2.3 PARALLEL CIRCUITS

The parallel circuit has more than one path for the electron flow. That is, in
a parallel circuit the electrons can follow two or more paths at the same time.
Electric devices (loads) are arranged in the circuit so that each is connected
to both supply voltage conductors.

Parallel circuits are common in the industry because most loads used
operate from line voltage. **Line voltage** is the voltage supplied to the equipment from the main power source of a structure and typically has a value of
110 volts. The parallel circuit allows the line voltage to reach all the electric
loads connected in parallel, as indicated in figure 2.9. Note in figure 2.9 that
each load in the circuit is supplied by the line voltage of 110 volts.

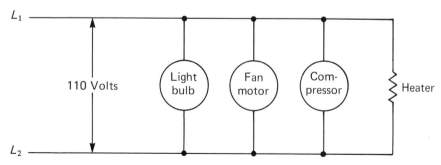

FIGURE 2.9. *Parallel circuit with four components; each component
is supplied with 110 volts of line voltage.*

Applications

Parallel circuits are used in the industry to supply the correct line voltage to several different circuits in a control system. Figure 2.10 shows a control system with several circuits in parallel being fed line voltage. There are many different paths for electron flow in this parallel hookup. Each circuit that is connected from L_1 to L_2 on the wiring diagram is in parallel with all the others and is being supplied with the line voltage.

Parallel circuits are used in all power wiring that supplies the loads of heating, cooling, or refrigeration systems. The electric loads of a system must be connected to the power supply separately or in a parallel circuit to supply the load with the full line voltage.

Calculations for Current, Resistance, and Voltage

There will be few occasions when field personnel are required to make calculations for a parallel circuit. This is usually done by the designer of the equipment. However, field personnel should be familiar with the basic concepts and rules of parallel circuits.

The current draw in a parallel circuit is determined for each part of the

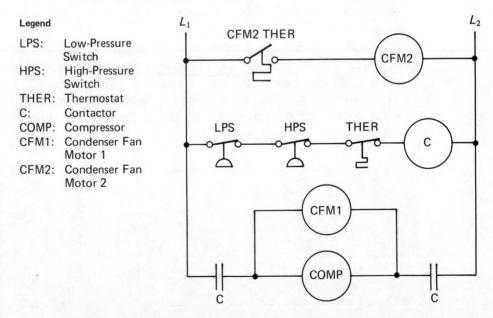

Legend

LPS: Low-Pressure Switch
HPS: High-Pressure Switch
THER: Thermostat
C: Contactor
COMP: Compressor
CFM1: Condenser Fan Motor 1
CFM2: Condenser Fan Motor 2

FIGURE 2.10. Control system with several circuits (parallel); each circuit is supplied by the line voltage.

circuit, depending on the resistance of that portion of the circuit. The total current draw of the entire parallel circuit is the sum of the currents in the individual sections of the parallel circuit. The current in each individual circuit can be calculated by using Ohm's law when the resistance and voltage are known. The total ampere draw of a parallel circuit is given by the following equation:

$$I_t = I_1 + I_2 + I_3 + I_4 + \cdots$$

The resistance of a parallel circuit gets smaller as more resistances are added to the circuit. The total resistance of a parallel circuit cannot be obtained by taking the sum of all the resistances. It is calculated by the following formula if two resistances are used:

$$R_t = \frac{R_1 \times R_2}{R_1 + R_2}$$

If three or more resistances are located in the circuit, the reciprocal of the total resistance is the sum of the reciprocals of all the resistances (the reciprocal of a number is one divided by that number. The following formula is used to calculate the resistance of a parallel circuit with more than two resistances:

$$\frac{1}{R_t} = \frac{1}{R_1} + \frac{1}{R_2} + \frac{1}{R_3} + \frac{1}{R_4} + \ldots$$

The voltage drop in a parallel circuit is the line voltage being supplied to the load. In other words, in a parallel circuit each load uses the total voltage being supplied to the load. For example, if 110 volts are supplied to a load, it will use the total 110 volts. The voltage being applied to each of the four components in figure 2.9 is the same and is given by the following equation:

$$E_t = E_1 = E_2 = E_3 = E_4$$

Ohm's law can be used to calculate voltage, amperage, or resistance if the other two values are known. You can use Ohm's law to determine almost any condition in a parallel circuit, but pay careful attention to the individual sections of each complete circuit.

Example 1   What is the total current draw of the parallel circuit shown in figure 2.11?

Step 1: First calculate the current draw for each individual circuit by using Ohm's law in the form

$$I = \frac{E}{R}$$

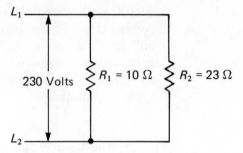

FIGURE 2.11. Parallel circuit for example 1.

$$I = \frac{E}{R}$$

Step 2: For I_1 substitute the given values for E and R_1 in the formula and solve:

$$I_1 = \frac{230}{10} = 23 \text{ amperes}$$

Step 3: For I_2 substitute the given values of that circuit in the formula and solve:

$$I_2 = \frac{230}{23} = 10 \text{ amperes}$$

Step 4: Use the formula $I_t = I_1 + I_2$.
Step 5: Substitute in the formula for I_t and solve:

$$I_t = I_1 + I_2$$
$$= 23 \text{ amperes} + 10 \text{ amperes} = 33 \text{ amperes}$$

Example 2   Find the total resistance of the parallel circuit in figure 2.11.

Step 1: Use the formula

$$R_t = \frac{R_1 \times R_2}{R_1 + R_2}$$

Step 2: Substitute the known values in the formula:

$$R_t = \frac{10 \times 23}{10 + 23} = 6.97 \text{ ohms}$$

Example 3   What is the resistance of a parallel circuit with resistances of 3 ohms, 6 ohms, and 12 ohms?

Step 1: Use the formula

$$\frac{1}{R_t} = \frac{1}{R_1} + \frac{1}{R_2} + \frac{1}{R_3}$$

Step 2: Substitute the known values in the formula:

$$\frac{1}{R_t} = \frac{1}{3} + \frac{1}{6} + \frac{1}{12}$$

Step 3: Mathematical computation of this formula is sometimes difficult. If you have any problem, consult your instructor.

$$\frac{1}{R_t} = \frac{4}{12} + \frac{2}{12} + \frac{1}{12} = \frac{7}{12}$$

$$R_t = 1.71 \text{ ohms}$$

2.4 SERIES-PARALLEL CIRCUITS

The series-parallel circuit is seldom used in individual circuits in the industry. It is more often seen on the full wiring layout of an air-conditioning, heating, or refrigeration unit. This type of electric circuit is a combination of the series and parallel circuits, as shown in figure 2.12. The series-parallel circuit is sometimes easier to understand when it contains only a few components. It becomes harder to understand when there are a large number of components.

The series-parallel circuit is often used to combine control circuits with circuits that supply power to loads, as shown in figure 2.13.

Any calculation of the values in a series-parallel circuit must be performed carefully because each portion of the circuit must be identified as series or parallel. Once the type of circuit has been determined, the calculations are made accordingly. The most common use of a series-parallel circuit in the industry is in the overall schematic diagram of a piece of equipment.

We now turn to a discussion of the basic principles of electric meters and the different kinds of meters used in the industry.

FIGURE 2.12. Series-parallel circuit with three components.

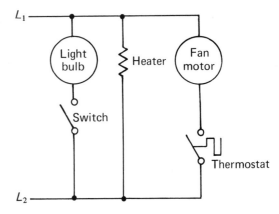

29

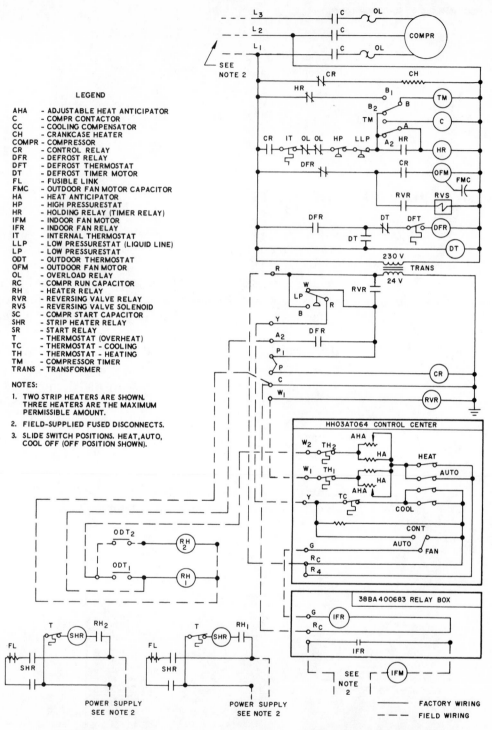

FIGURE 2.13. *Wiring diagram of an air-conditioning unit including a series-parallel circuit arrangement. Reproduced by permission of Carrier Corporation. © 1977 Carrier Corporation.*

2.5 ELECTRIC METERS

An electric meter is a device used to measure some electrical characteristic of a circuit. The most common types of electric meters are the voltmeter, the ammeter, and the ohmmeter.

Basic Principles

Most electric measuring instruments make use of the magnetic effect of electric current. When electrons flow through a conductor in an electric circuit, a magnetic field is created around that conductor, as shown in figure 2.14. This magnetic field is used to move the needle of a meter a certain distance, which represents the amount of the characteristic (volts, ohms, or amperes) being measured. The stronger the magnetic field, the larger the movement of the needle. The weaker the magnetic field, the smaller the movement of the needle.

If a compass is suspended next to a conductor that is not carrying an electron flow, the compass reacts only with the magnetic field of the earth and there is no other movement, as shown in figure 2.15. However, when electrons flow through that same conductor, the compass needle swings in line with the conductor's magnetic field, as shown in figure 2.16. The mechanical movement of the needle is caused by the magnetic field produced by the electron flow through the conductor. The larger the current flow, the stronger the magnetic field produced and the greater the needle movement on the scale. This simple principle is the basis of the meter movement in most electric meters.

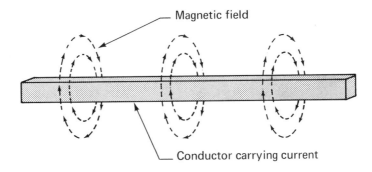

FIGURE 2.14. Magnetic field produced around a conductor when current is flowing through the conductor.

FIGURE 2.15. *When there is no current flow, the compass reacts only to the magnetic field of the earth.*

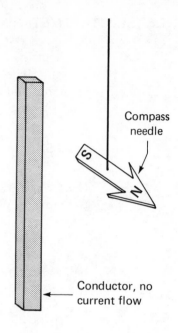

Compass
needle

Conductor, no
current flow

FIGURE 2.16. *When a current flows through a conductor, the compass needle swings in line with the conductor's magnetic field.*

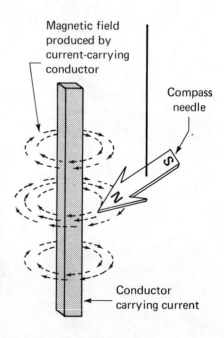

Magnetic field
produced by
current-carrying
conductor

Compass
needle

Conductor
carrying current

Differences Among Meters

The differences among the various electric meters is not in the meter movement but in the internal circuits of the meters and how the magnetic field is
created. To have electron flow in an electric circuit, an electric load must be
present. This electron flow is somewhat different for the clamp-on ammeter,
voltmeter, and ohmmeter. The clamp-on meter picks up the magnetic field
through a set of laminated jaws on the meter. The voltmeter uses a resistor
as a load to obtain an electron flow. The ohmmeter has its own power supply and uses a resistor for a load to produce a magnetic field. All three
meters use the same meter movement. Their methods of loads and power
supplies are varied to attain the needle movement for the reading.

Meters may be made in a combination and mounted in one case, or they
may be completely separate. Figure 2.17 shows a clamp-on ammeter and a
volt-ohm meter. The volt-ohm meter, as its name suggests, is used to measure
either voltage or resistance, depending on the scale selected.

The meter movement that is created by the magnetic field around a conductor is reflected by the movement of the needle on a scale of an electric
meter. This scale is usually broken down into several basic scales that have
different ranges for voltage, amperage, and resistance. Some electric meters
have a selector switch: the meter movement is shown on a certain point of
the scale but the scale to be used must be determined by the person using
the meter.

It is important for service personnel to understand the operation of an
electric meter because electric meters play a role in almost every part of the
industry, from the salesperson to the service mechanic. For example, the

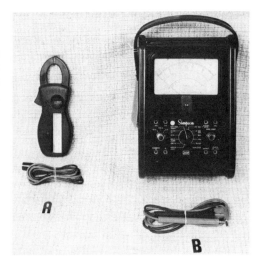

FIGURE 2.17. Clamp-on ammeter with
volt-ohm functions (a) and volt-ohm
meter (b). Photos courtesy of Amprobe
Instrument and Simpson Electric Co.,
Elgin, Ill.

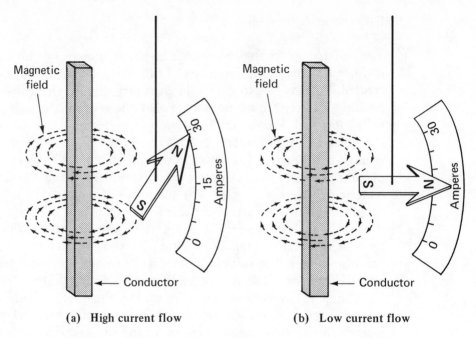

(a)   High current flow   (b)   Low current flow

*FIGURE 2.18.   Scale of an ammeter (a) Needle movement shows a
high current flow (b) Needle movement shows a low current flow.*

reading of the electric supply to a piece of equipment along with the current
draw gives the installation mechanic a good indication of whether the equipment is operating properly. Electric meters are essential for the service mechanic when called on to troubleshoot a piece of equipment. With the cost
of electric meters now at an all-time low, no installation or service mechanic
can afford not to take advantage of these valuable tools.

2.6 AMMETERS

The ammeter uses the basic meter movement discussed in section 2.5. The
strength of the magnetic field determines the distance that the needle of the
meter moves. Figure 2.18(a) shows the magnetic field and the meter movement when there is a high current flow through the conductor. Figure 2.18(b)
shows the magnetic field and the meter movement when there is a low current flow in the conductor. The larger the current flow, the stronger the
magnetic field grows and the greater the needle movement on the scale.

The ammeter measures current flow in an electric circuit. There are
basically two types of ammeters used in the industry today: the clamp-on

ammeter and the in-line ammeter. The clamp-on ammeter is the most popular because it is the easiest to use. You simply clamp the jaws of the meter around the conductor feeding power to the load that is producing the current draw. The clamp-on ammeter is shown in figure 2.19.

The in-line ammeter is used mostly on electric testing panels as a built-in component. It is also made with specially prepared leads to enable the service technician to use it as a plug-in unit. An in-line ammeter is shown in figure 2.20. The in-line ammeter must be connected in series with the load that is producing the current draw to give the correct reading. In most cases it is not practical to break a conductor and insert the in-line ammeter to obtain an amperage reading. Therefore the clamp-on ammeter is usually used.

The clamp-on ammeter is easy to use. Just follow a few simple rules. The jaws of the clamp-on ammeter are clamped around the conductor that is supplying a load or circuit. The magnetic field created by the current flowing through the wire is picked up by the jaws of the ammeter and funneled into the internal connection of the meter.

Never clamp the jaws of the meter around two wires to obtain an ampere reading. If the current flows in the wires are opposite, as they often are, the meter will read zero because the current flows cancel each other out. If the current flows are not opposite, the meter will read the current draw in both conductors. In either case you will obtain an incorrect reading, which could cause you to incorrectly diagnose the problem.

When the ampere draw is small, you may have difficulty obtaining a true reading because of the small needle movement. The problem of small needle movement can be remedied by coiling the wire around the jaws of the meter. This allows the meter to pick up a larger current flow than is actually there.

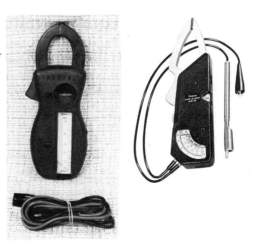

FIGURE 2.19. Clamp-on ammeters with volt-ohm functions. Photos courtesy of Amprobe Instrument and Simpson Electric Co., Elgin, Ill.

FIGURE 2.20. In-line ammeter. Photo courtesy of Simpson Electric Co., Elgin, Ill.

The meter will be more accurate because the current reading will fall in the midrange of the scale and can be easily read.

To obtain the correct ampere reading when this method is used, divide the ampere draw read by the number of loops going through the jaws of the meter. Figure 2.21 shows an ammeter with the conductor looped through the jaws three times. The correct reading can be obtained by dividing the meter reading, which is 4, by the number of loops through the jaws, which is 3. Thus the ampere draw of the load is 1.33 amperes. Remember: The meter reading should always be divided by the number of loops carried through the jaws of the meter.

The clamp-on ammeter is one of the most valuable tools that industry personnel can carry. When you are installing a system, knowing the ampere draw of the equipment tells you if the unit is operating properly. You can also detect many electric circuit problems with a clamp-on ammeter. The ammeter can be purchased with other meters built into it. For example, a

FIGURE 2.21. Taking a low-ampere reading with an ammeter by looping the wire through the jaws.

single clamp-on ammeter can be purchased that will read amperage, voltage, and resistance.

2.7 VOLTMETERS

The voltmeter is used to measure the amount of electromotive force available to a circuit or load. This is an important factor to heating, cooling, and refrigeration personnel because there is a wide range of voltages presently used in this country.

Voltmeters range from very simple to very complex instruments containing many scales. The simplest voltmeter available is a small inexpensive one capable of distinguishing only between 110 and 230 volts (see figure 2.22). Several manufacturers build simple voltmeters that can only read voltages, but these are becoming increasingly difficult to obtain. More common is the volt-ohm meter, which reads both voltage and resistance. These are available in many forms and service personnel should follow the instructions for the particular model being used. The common volt-ohm meter has three voltage scales and several voltage ranges. Some meters also have a high-voltage jack. Figure 2.23 shows three common types of volt-ohm meters.

The voltmeter is designed much like the ammeter, but a resistor is added to the circuit to prevent a direct short and allow electrons to flow in the meter. The voltmeter uses two leads that are connected to jacks that lead to the internal wiring. The two leads must touch or be connected to the conductors supplying the load or the circuit that transfers the electromotive force to the meter in order to obtain a reading. The electrons flow through the leads into the meter through a resistor with a known ohm rating. The

FIGURE 2.22.   Simple voltmeter. Photo courtesy of Knopp Inc.

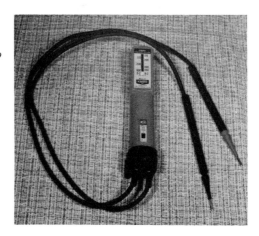

FIGURE 2.23.   Three volt-ohm meters commonly used in the industry. Photos courtesy of Amprobe Instrument, Simpson Electric Co., Elgin, Ill., and Triplett Corp.

greater the voltage carried into the meter, the greater the magnetic field and the greater the needle movement.

When you do not know the voltage available to the equipment being worked on, you should start with the highest scale on the meter. Then change the meter setting until the needle falls in the midrange of the voltage scale. Never abuse a voltmeter by attempting to read a higher voltage than the range of the meter.

The voltmeter is almost a necessity for field personnel who have anything to do with the electrical section of equipment or with the installation or servicing of equipment. No heating, cooling, or refrigeration equipment should operate at an unsafe voltage, that is, voltage that is either too low or too high. All equipment is designed to operate at a voltage of 10 percent above or below the rating of the equipment. But in some cases the voltage may actually be more than the allowable figures. So field personnel should always check the supply voltage. Installation mechanics have not completed their installation unless they have checked the voltage available to the equipment. Service mechanics are often required to check the voltages to equipment as part of their troubleshooting job. In addition, the voltmeter can be used by service mechanics as a tool for diagnosing problems in the system. Thus the voltmeter is a must for proper installation and service.

2.8 OHMMETERS

The ohmmeter is used to determine the operating condition of a component or a circuit. The ohmmeter can be used to find an open circuit, an open com-

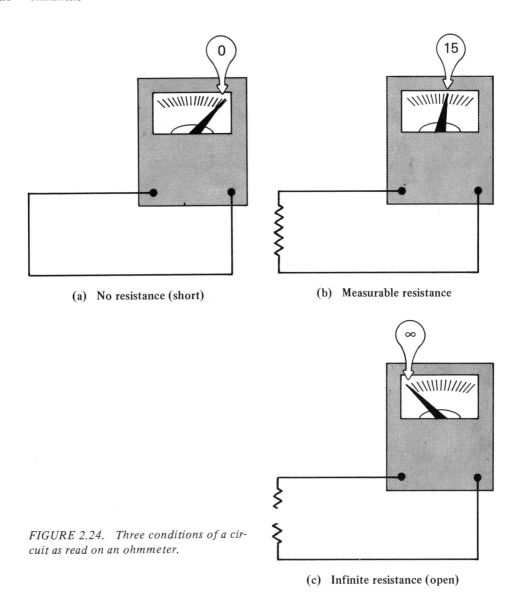

(a) No resistance (short)   (b) Measurable resistance

*FIGURE 2.24.   Three conditions of a cir-
cuit as read on an ohmmeter.*

(c) Infinite resistance (open)

ponent, or a direct short in a circuit or component. It can also be used to
measure the actual resistance of a circuit or a component.

The word **continuity** is used many times when referring to the use of
ohmmeters. Continuity means that a particular circuit or component has a
complete path for current to follow. An open component or circuit means
that there is no resistance or infinite resistance in the circuit. The term **measurable resistance** means the actual resistance that is measured with the ohmmeter. Figure 2.24 shows the three conditions as they might appear on the
scale of an ohmmeter.

The ohmmeter is a valuable tool for diagnosing and correcting problems in electric circuits. In the industry there are many electric devices and circuits that must be checked. The ohmmeter provides an easy method for checking circuits for opens (i.e., open circuits) and shorts (i.e., short circuits) and for measuring resistance.

An open circuit causes no noticeable needle movement in an ohmmeter because there is not a complete circuit. For example, an open (circuit) could occur in a blown fuse, a motor winding (the internal portion of a motor), and any condition where the electric circuit does not have a complete path for electrons to follow.

A direct short in an electric device or circuit causes problems because it means that two legs of the electric power wiring are touching, which causes an overload. In many cases a direct short means that the wiring of the component is connected in some fashion. A switch that is closed is considered to be a short, but without this type of short, no heating, cooling, or refrigeration system would operate properly.

In many cases you will have to measure the resistance of a component to ensure that the component is in good operating condition. Most manufacturers make available to service personnel the exact ohmic value of motor windings and other components in the system.

The meter movement of an ohmmeter is designed and built for a very low current that is available from its own power source, usually a battery. Figure 2.25 shows the internal wiring of an ohmmeter. The ohmmeter works much like the ammeter and the voltmeter except for the small current that is supplied from the internal power source. The ohmmeter also uses a magnetic field to move the needle, but the magnetic field is created by a self-contained power source in the meter. The two leads of the ohmmeter are connected to

FIGURE 2.25.   The internal wiring of an ohmmeter.

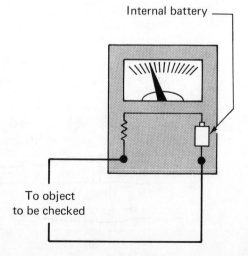

Internal battery

To object
to be checked

the internal circuit of the meter, which contains a resistance and the power source. The amount of current that the small battery can push through the device being tested indicates the resistance of that device and determines the needle movement.

Due to the low current that an ohmmeter is built to carry, it should never be connected to a circuit or device that is being operated. The function of the ohmmeter is merely to read the resistance of a device or circuit. Fortunately, almost all ohmmeters are equipped with some type of overload protection that will protect the internal circuits of the meter if they are subject to line voltage.

There are many types and designs of ohmmeters available. In many cases the ohmmeter and voltmeter are combined in a dual-purpose meter. Some manufacturers build and market combinations of voltmeters, ohmmeters, and ammeters that are fairly inexpensive. On this three-purpose meter the ohmmeter is a low-range ohmmeter and cannot be used for many of the jobs that a service mechanic must do. The more expensive volt-ohm meters are more accurate and cover all ohm ranges. These meters will usually have at least three ohm scales (usually $R \times 1$, $R \times 100$, and $R \times 10,000$). Some have more ranges. These additional ranges are useful for some troubleshooting operations. Figure 2.23 shows several different types of ohmmeters used today.

SUMMARY

The three types of electric circuits used in the industry are series, parallel, and series-parallel. The series circuit has only one path for electron flow. The most common type of electric circuit used in the industry is the parallel circuit, which has more than one path for electron flow. This type of electric circuit allows line voltage to reach all electric loads. The series-parallel circuit is a combination of the series and parallel circuits.

The series and parallel circuits have different relationships among the voltage, amperage, and resistance. In a series circuit the voltage is split among the electric loads. The current in the series circuit is equal in all parts of the circuit. The sum of all the resistances in series is the total resistance of the circuit. In a parallel circuit the voltage in all parts of the circuit is equal. The sum of each current flow in the circuit is the total current. The reciprocal of the total resistance is the sum of the reciprocals of all the resistances.

Electricity plays an important part in the industry. Almost all equipment has some type of electric control system, even if it is powered by some other means. Thus it is extremely important for industry personnel to be familiar with the basic electric meters. About 85 percent of the problems with equipment or systems turn out to be electrical, which shows that electric meters are important.

The ammeter is used to measure the current flow in an electric circuit. There are two types of ammeters used in the industry today. The clamp-on ammeter is by far the most frequently used. With this type of ammeter it is only necessary to clamp the jaws around the conductor feeding the circuit or load and read the amperage. The in-line ammeter must be put in series with the load or circuit to read the amperage. Because of the time required to do this, the in-line ammeter is seldom used.

The voltmeter is used to measure the voltage of an electric circuit. An ohmmeter is used to measure the resistance in an electric circuit. These two electric meters are often built in combination as the volt-ohm meter. It is also possible to get all three meters in combination.

QUESTIONS

1. What is a series circuit?

2. What is a control circuit?

3. Why are series circuits used for most control circuits?

4. What is voltage drop?

5. What is a parallel circuit?

6. Why are parallel circuits used to feed electric energy to loads?

7. Compare the calculations for voltage, amperage, and resistance in a series and in a parallel circuit.

8. What is a series-parallel circuit?

9. What are the requirements for an electric circuit?

10. What are the three most common electric meters used in the industry?

11. Why should industry personnel be proficient at using electric meters?

12. Almost all electric meters make use of the _____.

13. How does an ammeter work? p. 35

14. What are the two types of ammeters? Which type is more commonly used?

15. What is the result on the meter reading of clamping the jaws of a clamp-on ammeter around two wires?

16. How can very small ampere draws be measured with a clamp-on ammeter?

17. How does a voltmeter operate? p. 37

18. Air-conditioning or refrigeration equipment can operate properly at _10_ percent above or below its rated voltage.

19. If a service mechanic has no idea of the voltage available to a unit, what procedure should be followed on reading the voltage?

20. What precaution should be taken when using an ohmmeter?

42

21. What does the term "continuity" mean?

22. What is a short circuit?

23. What is an open circuit?

24. What factors should be considered when purchasing an electric meter?

25. What is the reaction of an ohmmeter to a measurable resistance, a short circuit, and an open circuit?

26. An in-line ammeter must be connected to the circuit in _____.

27. A voltmeter is connected to a circuit in _____.

28. State the differences among the ohmmeter, the voltmeter, and the ammeter as far as their internal circuitry is concerned.

29. What is the resistance of a series circuit with resistances of 2 ohms, 4 ohms, 6 ohms, and 10 ohms?

30. What is the resistance of a parallel circuit with resistances of 10 ohms and 20 ohms?

31. What is the total ampere draw of a parallel circuit with ampere readings of 2 A, 7 A, and 12 A?

32. What is the voltage of a series circuit with four voltage drops of 30 volts each?

3

Components, Symbols, and Circuitry of Air-Conditioning Wiring Diagrams

INTRODUCTION

Because of the complexity of today's air-conditioning, heating, and refrigeration systems, industry personnel should be able to read and interpret all kinds of wiring diagrams. Electric wiring diagrams contain a wealth of information about the electrical installation and operation of the equipment. The installation mechanic depends on the wiring diagram for the correct installation of the wiring to the unit. The service mechanic uses the electrical diagrams as a guide in troubleshooting the electric system of a unit.

It would be impossible for wiring diagrams to be composed of photographs of various components of the equipment. They would be too large and in many cases too complex due to the number of wires that are carried to certain devices. Thus symbols are used in wiring diagrams to represent such system components as compressors, indoor fan motors, thermostats, pressure switches, and heaters. Industry personnel must be able to identify most symbols and know where to look up the remainder.

Most manufacturers use similar symbols for each type of electric component, although there are some minor differences in symbols between some major manufacturers. Thus a knowledge of the basic symbols is essential if you are to be successful in the industry.

FIGURE 3.1. Electric motor. Photo courtesy of Emerson Electric Co., Emerson Motor Div.

We begin our study with a discussion of the various types of electric loads found in the industry and the basic symbol used for each device.

3.1 LOADS

Loads are electric devices that consume electricity to do useful work. Loads are devices such as motors (figure 3.1), solenoids (figure 3.2), resistance heaters (figure 3.3), and other current-consuming devices. The size of loads varies from devices with a small current draw, such as a light bulb, a small fan motor, and solenoids, to large motors that could use upwards of 100 amperes.

Loads are the most important part of a heating, cooling, or refrigeration system because they do all the work in the system. Loads operate **compressors**, which compress and transfer refrigerant in a system. They operate fans,

FIGURE 3.2. A solenoid used to operate a contactor. Photo courtesy of Furnas Electric Co.

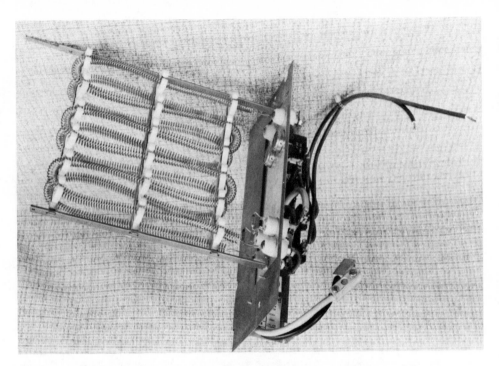

FIGURE 3.3.   A resistance heater. Reproduced by permission of Carrier Corporation. © 1977 Carrier Corporation.

which move air. They operate the solenoid part of a relay, which starts and stops loads. And loads operate other devices that perform useful work. Industry personnel should be able to recognize the common symbols for loads and know where to look up the symbols for little-used loads, because each electric wiring diagram is composed of symbols and their interconnecting wires.

In the following paragraphs we will take a close look at several of the different kinds of loads used in the industry.

Motors

A **motor** is an electric device that consumes electric energy to rotate a device in an electric system. Motors are used in the industry to rotate devices such as a compressor (figure 3.4), a condenser fan motor (figure 3.5), pumps (figure 3.6), and other units that require rotating movement. Motors are the largest and most important loads in heating, cooling, and refrigeration systems.

The symbol shown in figure 3.7(a) is the most common symbol used to

FIGURE 3.4. Compressor.

represent motors. The other symbol used to represent motors is shown in figure 3.7(b).

A letter designation tells you what purpose the motor serves in the system. Figure 3.8 shows several symbolic representations of different uses of motors. Careful attention should be given to symbols representing motors because in some cases a motor has an internal overload, as shown in figure 3.8(e).

FIGURE 3.5.   A condenser fan motor on a small air conditioner. Photo courtesy of Fedders Corp.

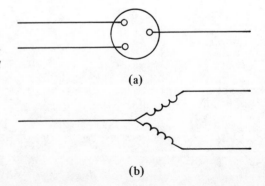

FIGURE 3.6. A pump used as a condenser water pump. Photo courtesy of John L. Underwood Co., Inc.

Solenoids

The **solenoid** is a device that creates a magnetic field when energized and causes some action to an electric component such as a relay or valve. Two common solenoids used to operate a relay and a contactor are shown in figure 3.9. The solenoid is considered a load because it consumes electricity to do useful work.

Solenoids are devices that open and close to control some element in a system. Solenoid valves are valves that open and close, stopping or starting a flow. Solenoid coils used in relays and contactors will be discussed later in this chapter. Some common solenoid valves are hot-gas solenoids, reversing-valve solenoids, and liquid-line solenoids. Figure 3.10 shows a solenoid valve, and a solenoid coil used on a solenoid valve with its symbol.

Heaters

Heaters are loads that are found in many systems and wiring diagrams. A heater takes electric energy and converts it to heat. In some cases electric resistance heaters are used to heat homes. Heaters might also be used to heat

FIGURE 3.7. Symbols for an electric motor (a) Most commonly used symbol (b) Alternate symbol.

(a)

(b)

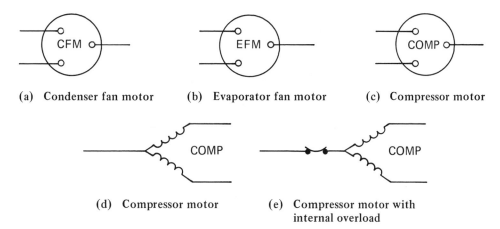

(a) Condenser fan motor   (b) Evaporator fan motor   (c) Compressor motor

(d) Compressor motor   (e) Compressor motor with
                                        internal overload

FIGURE 3.8.   Symbols representing some common uses of motors.

a small object or area. The symbol for all heaters is the same. Only a letter designation tells you specifically why the heater is used. Figure 3.11 shows the symbols used for heaters, along with some of the more common letter designations.

Signal Lights

A **signal light** is a light that is illuminated to denote a certain condition in a system. The letter inside the signal light symbol denotes the color of the signal light, as shown in figure 3.12. Signal lights come in a variety of colors and are not limited to the colors shown in figure 3.12. A signal light is used to show that a piece of equipment is operating or that it is operating in an unsafe condition. Signal lights are usually energized when a piece of equipment or component is started.

FIGURE 3.9. Two solenoids used to operate a relay. Photo courtesy of Essex Group, Controls Div.

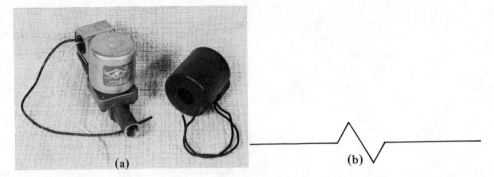

FIGURE 3.10.   A solenoid valve (a) and solenoid coil with symbol (b).
Photos courtesy of Jackes-Evans Mfg. Co.

3.2   CONTACTORS AND RELAYS

Contactors and **relays** are devices that open and close a set or sets of electric contacts by the action of a solenoid coil. The contactor or relay is composed of a solenoid and the contacts. A relay is shown in figure 3.13. A contactor is shown in figure 3.14. When the solenoid is energized, the contacts will open or close, depending on their original position (that is, if they were open they will close, and vice versa).

In an air-conditioning control system we must have some method of controlling loads. In most cases a relay or contactor is used. Relays and contactors are widely used in control systems. Thus it is essential that industry personnel be able to identify the symbols for relays and contactors.

The only difference between a relay and contactor is in the size of the device. A contactor is simply a large relay. Usually the devices are distinguished by their rated current flow. A contactor can carry 20 amperes or more. A relay is designed to carry less that 20 amperes. It is very likely for

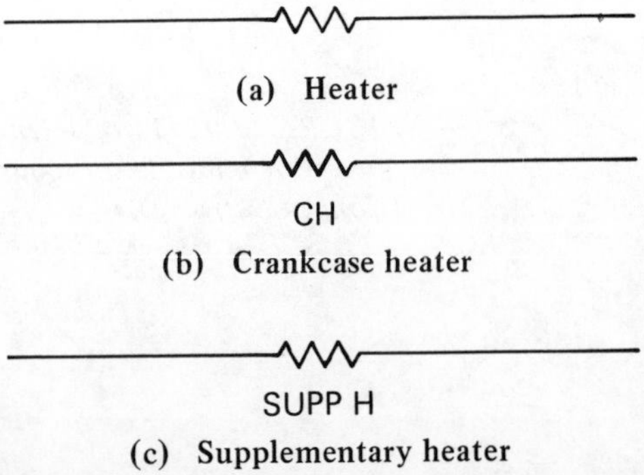

FIGURE 3.11. Symbols for common
uses of resistance heaters.

(a) Heater

CH

(b) Crankcase heater

SUPP H

(c) Supplementary heater

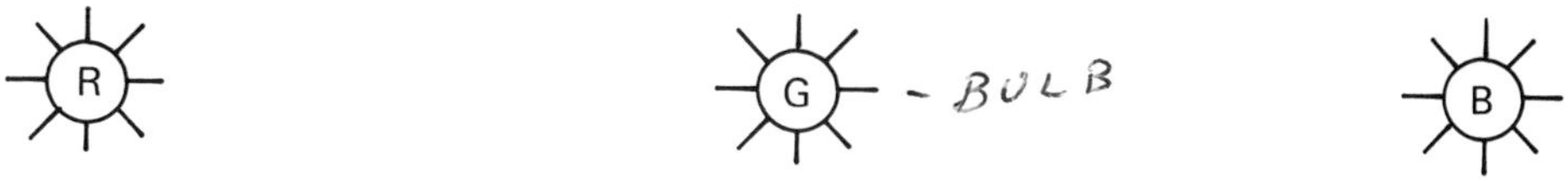

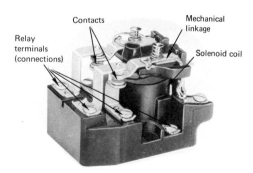

(a) Red   (b) Green   (c) Blue

FIGURE 3.12. Symbols for signal lights showing the color of the light.

contactors to be used where the ampere draw of a device is more than 20 amperes. However, a relay would rarely be used to carry over 20 amperes.

Contactors and relays play an important part in the control system of any air conditioner, refrigerator, or heater. For example, contactors and relays are used to stop and start different loads in a refrigeration system. Compressors in most air-conditioning systems are controlled by a contactor or magnetic starter. Relays can be used for pilot duty, that is, for controlling another relay or contactor. The most important fact to remember is that in most control systems there are many relays and at least one contactor. These relays or contactors always control some load.

Relays and contactors are composed of two parts: the contact and the coil, or solenoid. The contact makes the electrical connections. Figure 3.15 shows the symbol for a **pole**, or contact, of a relay or contactor. The term "pole" refers to one set of contacts. However, in some cases the relay or contact might have two or three poles, which means two or three sets of contacts. The coil or solenoid, the second part of the relay, is **energized** (voltage is supplied) and, through a magnetic field, closes the contact or contacts. Either symbol shown in figure 3.16 can be used to represent a relay or contactor coil. The symbol for the relay or contactor is the same if each has the same number of poles and if their purpose is basically the same, with the exception of the ampere rating of the device.

All symbols are usually shown in the **deenergized** position. This means

FIGURE 3.13. Relay. Photo courtesy of Potter & Brumfield Div., AMF Inc.

Solenoid

Terminals

FIGURE 3.14. Contactor. Photo courtesy of Furnas Electric Co.

that there is no electric potential to the coil of the device. Figure 3.15 shows a pole in the deenergized position.

The term **normally** refers to the position of a set of contacts when the device is deenergized. The contacts in figure 3.15 are called **normally open**. This means that they would close when the relay or contactor is energized. Shown in figure 3.17 is a relay with a set of **normally closed** contacts. These contacts would open if the relay was energized.

The terms "normally open," "normally closed," "energized," and "deenergized" are important in understanding relays and contactors on wiring diagrams. Figure 3.18(a) shows a relay with two normally open contacts and one normally closed contact in the deenergized position (with no voltage to the coil). Figure 3.18(b) shows the same contacts in the energized position (with voltage to the coil). In the deenergized position the current will not

FIGURE 3.15. Symbol for a normally open pole of a relay or contactor.

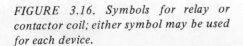

FIGURE 3.16. Symbols for relay or contactor coil; either symbol may be used for each device.

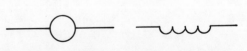

FIGURE 3.17. Normally closed set of contacts. Photo courtesy of Potter & Brumfield Div., AMF Inc.

flow through contacts 1 and 2, but current will flow through contact 3. In the energized position the current flow is through 1 and 2 but not through 3.

3.3 MAGNETIC STARTERS

A **magnetic starter** is the same type of device as a contactor in terms of the ampere rating of the device. But the magnetic starter has a means of overload protection in it, whereas the contactor has none. Figure 3.19 shows a picture of the magnetic starter and its symbol. The principle of operation of the magnetic starter will be covered in chapter 7.

3.4 SWITCHES

An electric **switch** is a device that opens and closes to control some load in an electric circuit. Electric switches can be opened and closed by tempera-

FIGURE 3.18. Symbols showing relays deenergized and energized.

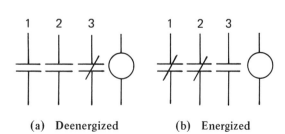

(a) Deenergized   (b) Energized

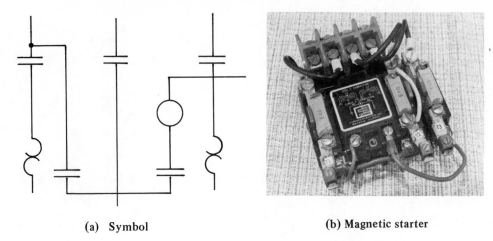

(a) Symbol

(b) Magnetic starter

FIGURE 3.19.   Magnetic starter. Photo courtesy of Furnas Electric Co.

ture, pressure, humidity, flow, or by some manual means. You must become familiar with the symbols used for switches because in most cases they control the loads in the system. The symbol will also indicate what is initiating the action of the switch.

A manually operated switch is a switch that is opened and closed by manual force. Figure 3.20 shows a simple manually operated switch. The poles of a manual switch are the number of contacts that are included in the switch. The throw indicates how the switch may be operated. For example, a single-pole–single-throw switch has one set of contacts and two positions: an open and a closed position, as shown in figure 3.20. A double-pole–double-throw switch has two sets of contacts and two positions, as shown in

FIGURE 3.20. Simple manual switch; single-pole–single-throw switch. Photo courtesy of Eagle Electric Mfg. Co., Inc.

FIGURE 3.21. Double-pole–double-
throw manual switch. Photo courtesy of
Eagle Electric Mfg. Co., Inc.

figure 3.21. Symbols for these two switches and for two other basic types of manual switches are shown in figure 3.22.

There are other types of manual switches used in the industry. The **disconnect switch** is used to open and close the main power source to a piece of equipment or load. Figure 3.23 shows a three-pole disconnect switch and its symbol. The **push-button switch,** as shown in figure 3.24, is a switch used to open and close a set of contacts by pressing a button. The symbols for the normally closed and the normally open push-button switches are also shown in figure 3.24.

The most important type of switch in a control system is the mechanically operated switch. **Thermostats** are mechanically operated switches used in almost all control systems. Thermostats are said to be mechanically operated because the temperature-sensing element moves a set of contacts by a mechanical linkage. Thermostats are designed for heating, cooling, or both. The cooling thermostat is designed to close on a temperature rise and open on a temperature fall. The heating thermostat is designed to open on a temperature rise and close on a temperature fall. The symbols for these two

FIGURE   3.22. Symbols   for   manual
switches.

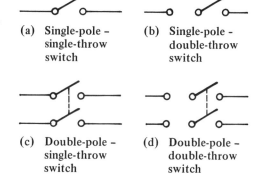

(a) Single-pole –
single-throw
switch

(b) Single-pole –
double-throw
switch

(c) Double-pole –
single-throw
switch

(d) Double-pole –
double-throw
switch

(a) Switch

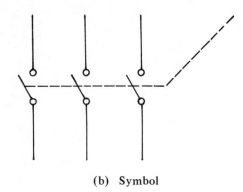

(b) Symbol

FIGURE 3.23.   *Three-pole disconnect switch. Photo courtesy of Gould I-T-E Electrical Products.*

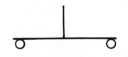

(b) Symbol for
    normally closed
    push-button
    switch

(c) Symbol for
    normally open
    push-button
    switch

FIGURE 3.24.   *Push-button switch. Photo courtesy of Gould I-T-E Electrical Products.*

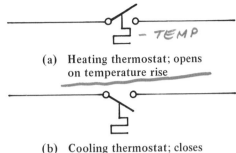

(a) Heating thermostat; opens
     on temperature rise

(b) Cooling thermostat; closes
     on temperature rise

FIGURE 3.25. *Symbols for heating and cooling thermostats.*

types of thermostats, shown in figure 3.25, indicate their function. Figure 3.26 shows a modern thermostat.

Pressure switches are used for different functions in modern control circuits. The purpose of the pressure switch determines whether it opens or closes on a rise or fall in pressure. The pressure range of the switch is not part of the symbolic representation. Figure 3.27 shows the symbols for pressure switches. Letter designations in the symbols often denote the pressure

FIGURE 3.26. *Thermostat. Photo courtesy of Honeywell, Inc.*

Mercury
Bulb

Bimetal

Temperature
Setting

Thermometer

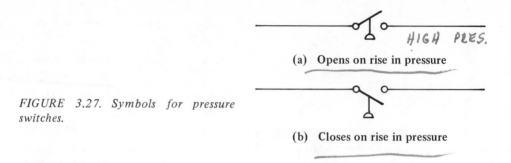

(a) **Opens on rise in pressure**

FIGURE 3.27. Symbols for pressure switches.

(b) **Closes on rise in pressure**

ranges and purposes of the switches. Figure 3.28 shows some common pressure switches used in the industry.

3.5 SAFETY DEVICES

Safety devices are very important in today's modern systems. Components are becoming more expensive each year. Thus it is vital that these components be protected from adverse conditions such as low voltage, high ampere draw, and overheating. It is for this reason that you should become familiar with symbols for safety devices. Overloads and safety devices are sometimes a combination of a load and a switch. They differ from the relay in their purpose and overall design.

All motors are designed to operate on a certain current draw. If for some reason this rating is exceeded, the motor must be cut off immediately to

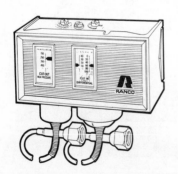
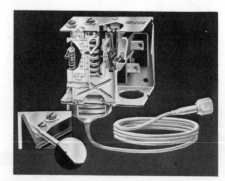
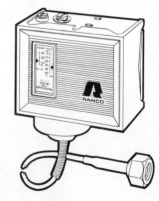

FIGURE 3.28. Some common pressure switches. Photos courtesy of Ranco Controls Division.

FIGURE 3.29. Symbols for fuses.

prevent damage and possible destruction of the component. A burned out motor is often caused by a malfunction in the safety devices.

The **fuse** is the simplest type of overload device. The fuse is effective against a large overload, but it is less effective against small overloads. The fuse is nothing more than a piece of metal designed to carry a certain load. Any higher load will cause the fuse to break the circuit. Figure 3.29 shows two symbols for a fuse. Figure 3.30 shows some common fuses in use today.

The second type of overload device is designed to protect the motor against small and large overloads. They are divided into two categories: thermal and magnetic. The **thermal overload** is operated by heat and the **magnetic overload** is operated by magnetism, which is directly proportional to the current draw.

The thermal overload can be a **pilot duty** device, which breaks the control circuit and locks the motor out. The pilot duty types of overloads are most common on motors larger than 3 horsepower. Or the thermal overload can be a line voltage device, which breaks the power line to the component being protected.

The bimetal element is the simplest of the thermal overloads. When it gets warm, it warps to open the circuit, as shown symbolically in figure 3.31. Some bimetal elements are furnished with heaters, as shown symbolically in figure 3.32. The heater allows the bimetal disc to react to an overload quicker because the current flow is proportional to heat.

The thermal overload relay, whose symbols are shown in figure 3.33, is a simple device with a thermal element and a switch that opens on a rise in temperature.

The magnetic overload symbol is the same as the symbol for a relay with one normally closed contact. The current flow is relayed to the overload

FIGURE 3.30. Some common fuses. Photo courtesy of The Chase Shawmut Co.

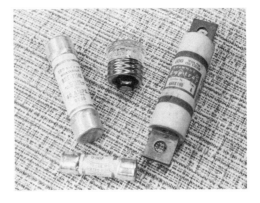

FIGURE 3.31. Symbols for bimetal overload.

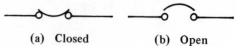

(a) Closed   (b) Open

FIGURE 3.32. Symbol for three-wire bimetal overload.

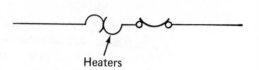

Heaters

coil. Since current flow is proportional to the strength of the magnetic field, the relay can be designed to energize only on a high current draw. Figure 3.34 shows a magnetic overload and its symbol. The letter designation of this device will distinguish between the magnetic overload and the common relay.

3.6 TRANSFORMERS

The **transformer** decreases or increases the incoming voltage to a desired voltage. In most air-conditioning control circuits, it is not practical to pull large wires for a long distance. Therefore a 24-volt control circuit, which is safer, less expensive, and a better method of control, is used. Figure 3.35 shows a transformer and its symbol. The voltage is also given with the symbol in some cases.

3.7 SCHEMATIC DIAGRAMS

Most modern heating, cooling, and refrigeration systems are becoming more complex with more controls and safety devices. Advances in controls and control systems make it a necessity for you to be able to read **schematic diagrams**. If you are able to read schematic diagrams, you will know what the unit should be doing.

FIGURE 3.33. Symbols for a thermal overload relay.

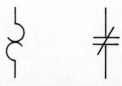

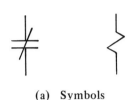 

(a) Symbols   (b) Magnetic overload device

FIGURE 3.34. Magnetic overload. Photo courtesy of Heinemann Electric Co.

The schematic diagram is the most useful and easiest to follow of any electric diagram. The schematic diagram tells how, when, and why a system works as it does. In most cases service technicians use schematic diagrams to troubleshoot control systems. The schematic wiring diagram includes the symbols and the line representations so that the user can easily identify loads and switches along with the circuits.

All electric circuits contain a source of electrons, a device that uses electron flow, and a path for the electrons to follow. In most cases the source of electrons is an alternating current voltage supply. The device using the electron flow is a motor, heater, relay coil, or any other load device. The path for the electrons to follow is a wire or any type of conductor.

(a) Symbol   (b) Transformer

FIGURE 3.35. Transformer. Photo courtesy of Penn Division, Johnson Controls, Inc.

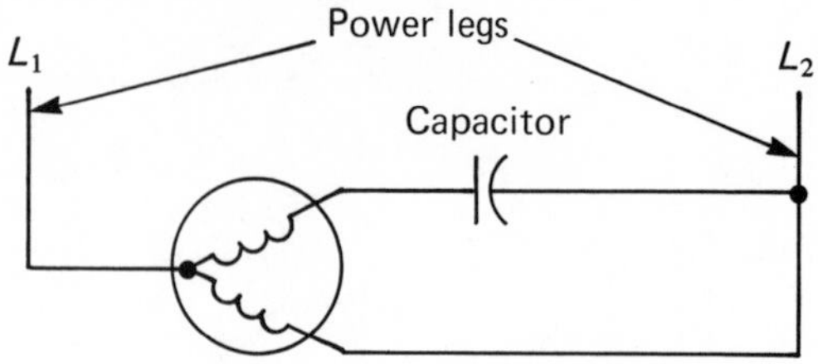

FIGURE 3.36. *Schematic diagram of a complete circuit.*

In the schematic diagram the source of electrons, the power supply, is represented by two lines drawn downward and listed as L_1 and L_2, as shown in figure 3.36. There is a potential difference of 230 volts between L_1 and L_2. If a path is created between L_1 and L_2, current will flow.

All electric loads in the unit are placed between L_1 and L_2 along with the switches controlling the load. Figure 3.37 shows a complete circuit in schematic form with a compressor and the switch that controls it. When the switch is closed, the motor will run. In figure 3.37 the source of electrons is from L_1 and L_2, the path is the connecting wire, and the device using the electron flow is the compressor. The compressor operates when the switch is closed.

Figure 3.38 shows a full schematic diagram similar to the diagrams you will be using on the job. All schematic diagrams are broken down into a circuit-by-circuit arrangement. Almost all schematic diagrams contain a legend that cross-references the components and their letter designation to the name of the component. Look for the legend in figure 3.38.

3.8 READING SIMPLE SCHEMATIC DIAGRAMS

Schematic diagrams are used throughout the industry to show a systematic layout of a control system. Schematic diagrams range from simple to very complex, depending on the use. Most residential systems use a simple sche-

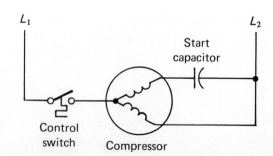

FIGURE 3.37. *Schematic diagram of a complete circuit with a control.*

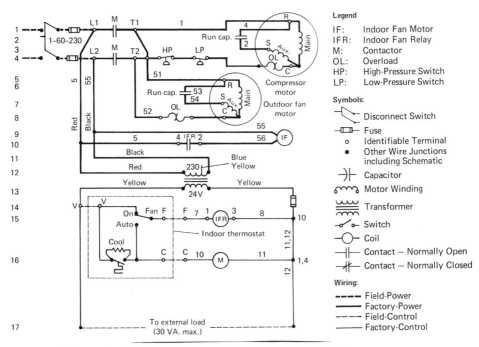

FIGURE 3.38. Complete schematic diagram for small packaged unit. Diagram courtesy of Westinghouse Electric Corp., Central Residential Air Conditioning Div.

matic with few components. A large commercial and industrial system usually has a complex diagram, which is harder to read and understand. However, the layout arrangement of a schematic diagram is the same whether the system is residential or commercial and industrial. This section will cover two basic schematics, circuit by circuit.

Schematic Wiring Diagram of a Simple Air Conditioner

The first schematic we will discuss is shown in figure 3.39. The source of electrons comes from L_1 and L_2. The potential difference of the control circuit in figure 3.39 is 230 volts. This diagram represents a simple air conditioner with line voltage controls. The indoor fan motor is controlled by a switch and must be on for the compressor circuit to operate.

Figure 3.40 shows the basic circuit of the indoor fan motor (IFM) operated with a double-pole–single-throw switch. Note that one pole is not used in the IFM circuit. When the switch is closed, the IFM will operate.

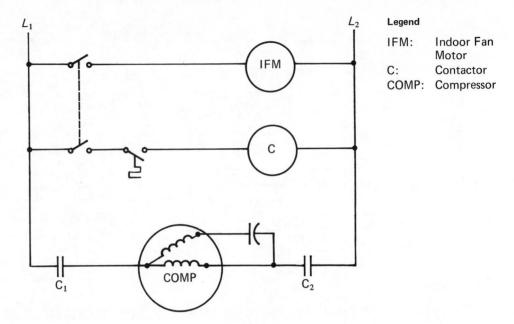

FIGURE 3.39. Schematic wiring diagram of a simple air conditioner.

Figure 3.41 shows the IFM circuit and the contactor circuit. The contactor coil circuit has two switches that control the circuit. The purpose of the manual double-pole–single-throw switch used in the contactor coil circuit is to keep the contactor coil from being energized except when the IFM is operating. The thermostat energizes the contactor coil to maintain the thermostat setting. When the contactor coil is energized, the C1 and C2 contacts of figure 3.39 close and start the compressor. If the thermostat becomes satisfied, the contactor is deenergized and the compressor is stopped.

FIGURE 3.40. Indoor fan motor circuit of the schematic diagram.

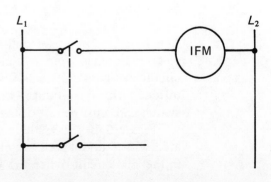

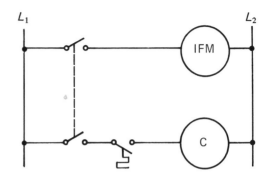

FIGURE 3.41. Indoor fan motor circuit with contactor coil circuit.

Schematic Diagram of a Packaged Air-Conditioning Unit

The next schematic we will discuss is that of a simple air-cooled package unit with a low-voltage control system. The complete schematic is shown in figure 3.42. A packaged unit is a unit that is built in one piece.

Almost all residential air-conditioning systems have a 24-volt control circuit (see figure 3.42) to operate the electric control components. When a 24-volt control system is incorporated, more components are required because a transformer must be used to reduce line voltage to the 24-volt low-voltage supply.

The schematic diagram of figure 3.42 is basically divided into two sections. The portion of the diagram above the transformer is the line voltage section. The portion below the transformer is the 24-volt control circuit. The only connection between the two circuits is the transformer and the fact that the contacts of the low-voltage relays are connected into the line portion of the control system.

The schematic for the low-voltage thermostat used on the unit is shown in figure 3.43. The thermostat is a combination of several switches along with the actual temperature-detecting portion of the control, which in this case is a cooling thermostat. The switch in the upper left of the thermostat diagram is a system switch. It cuts the total control system off; that is, it stops the entire system from operating. The fan switch allows the fan to be operated alone on either the "on" or the "manual" setting. When the fan switch is set on automatic, the fan operates only when the compressor operates.

The low-voltage control portion of the system opens and closes relays and contactors in the line voltage portion of the circuit. The low-voltage control circuit also operates the compressor and condenser fan motor. The contactor coil of the low-voltage system (see figure 3.42) is energized through the cooling thermostat, assuming that the system switch is in the "on" position. When the contactor coil is energized, its contacts close, starting the

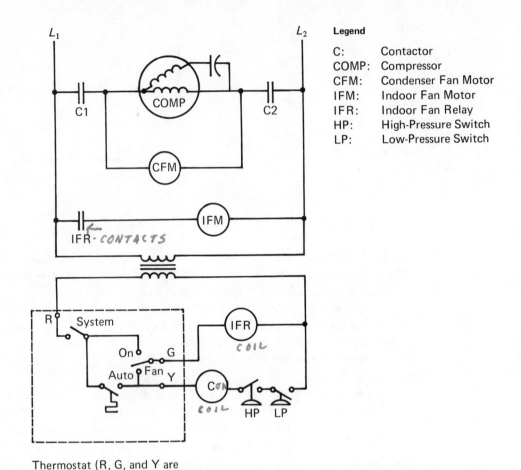

FIGURE 3.42.   Schematic diagram of a packaged air-conditioning unit.

compressor and condenser fan motor. The high-pressure and low-pressure switches in the contactor coil circuit will deenergize the contactor coil whenever the refrigeration system reaches an unsafe operating condition, be it high or low pressure in the refrigeration system.

The upper portion of the schematic diagram in figure 3.42 is divided into three separate circuits. These circuits are discussed individually in the following paragraphs.

Figure 3.44 shows the compressor and condenser fan motor circuit. The circuit contains the compressor and condenser fan motor. The compressor and condenser fan motor are stopped and started by the contacts of the contactor.

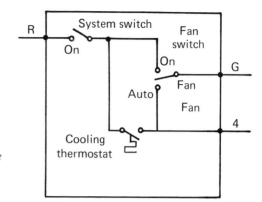

FIGURE 3.43. Schematic diagram of the 24-volt thermostat with all switches.

The indoor fan motor circuit is shown in figure 3.45. The indoor fan motor starts when the contacts on the indoor fan relay close.

Figure 3.46 shows the indoor fan relay coil circuit and the line voltage circuits of the indoor fan motor. The fan is operated by closing the indoor fan relay contacts. When the thermostat fan switch is in the "on" position, the indoor fan relay energizes and closes the IFR contacts, operating the indoor fan continuously until the fan switch is moved to the automatic position. When the fan switch is in the automatic position, the indoor fan relay coil is energized whenever the compressor is operating.

The following outline gives the sequence of operation of the air-conditioning unit. Refer to figure 3.42 as you read each step.

1. The contactor coil is energized by the cooling thermostat when the system switch is in the "on" position. When the contactor closes, the compressor and condenser fan motor start.
2. The fan motor operates continually with the fan switch in the "on" position because the IFR coil is energized, closing the contacts. If the fan switch on the thermostat is in the automatic position, the fan operates when the compressor is running.
3. The high-pressure and low-pressure switches are safety devices that deenergize the contactor if either of the switches open.

The schematic diagrams for residential and small commercial systems are

FIGURE 3.44. Schematic diagram of the compressor and condenser fan motor circuit.

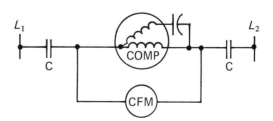

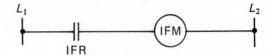

FIGURE 3.45. *Schematic diagram of the indoor fan motor circuit.*

usually fairly simple and easy to understand. It is important for service personnel to be able to interpret wiring diagrams so that they can service the equipment. A schematic wiring diagram can tell a service mechanic how and when a piece of equipment operates as it does.

3.9   READING ADVANCED SCHEMATIC DIAGRAMS

Schematic diagrams can become increasingly complex due to the addition of various components that are needed in control systems in commercial and industrial units. The service mechanic should be able to read all schematics in order to troubleshoot the control systems of all air-conditioning, heating, or refrigeration systems. The service mechanic should keep in mind that a schematic is broken down in a circuit-by-circuit arrangement. Once a service mechanic has realized this, a schematic diagram, no matter how complex, becomes readable. Experience in the industry leads service mechanics to become more proficient at reading all types of diagrams.

A schematic diagram of a large condensing unit is discussed in this section. The unit is energized by applying 230 volts to the condensing unit. Unlike the small residential units, the control circuit of this unit is completely line voltage. The unit operates on 208 volts-3 phase-60 cycles.

FIGURE 3.46. *Schematic diagram of the indoor fan motor and low-voltage circuits.*

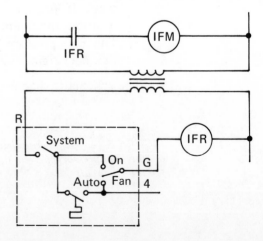

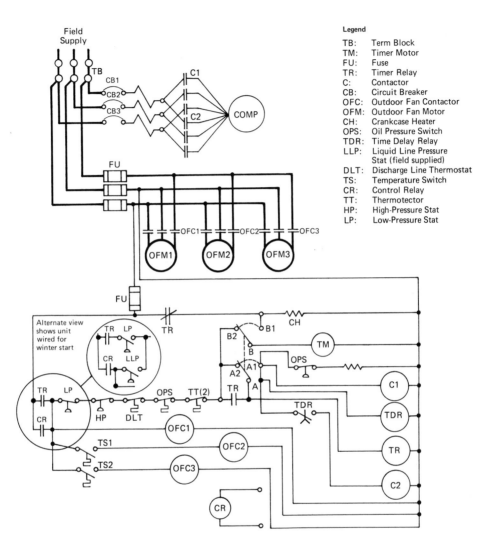

FIGURE 3.47.   Schematic diagram of a large commercial and industrial condensing unit. Reproduced by permission of Carrier Corporation. © 1977 Carrier Corporation.

The schematic diagram for the unit discussed in this section is shown in figure 3.47. The unit is started by the energization of the control relay. This relay could be energized by a thermostat, pressure switch, or any other controlling device. The diagram will be discussed in four sections: the timer and timer relay circuits, the compressor circuit, the condenser fan motor circuits, and the safety controls of the circuit.

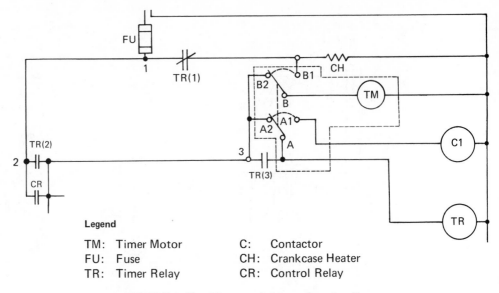

Legend

TM:	Timer Motor	C:	Contactor
FU:	Fuse	CH:	Crankcase Heater
TR:	Timer Relay	CR:	Control Relay

FIGURE 3.48.   Timer and timer relay circuits.

Timer and Timer Relay Circuits

Figure 3.48 shows the timer and timer relay circuits with the timer enclosed by a dashed line. This timer and timer relay arrangement is used by several manufacturers to protect the system from **short cycling** (rapid starting and stopping). The timer allows for a 5-minute delay from the time the unit shuts down to when it starts again.

The timer allows for a 5-minute delay from one cycle of the compressor to another cycle. The timer makes one complete revolution in a 5-minute period. The timer rotates 4 minutes and 45 seconds with the switches in the position from B to B1 and from A to A1. The timer runs because of the normally closed contacts of the timer relay TR(1) located in circuit 1. After the timer motor has run 4 minutes and 45 seconds, it changes the position of the timer contacts from A to A2 and from B to B2 and stops the timer motor. This has no effect on the system until the control relay contacts in circuit 3 are closed.

When the control relay CR is closed, the timer relay becomes energized by circuit 3 through A2 to A of the timer. This action in turn closes the timer relay contacts TR(2) and TR(3) and opens contact TR(1). Once the timer relay is energized, the timer motor runs an additional 15 seconds and changes positions from A to A1 and from B to B1, which energizes the contactor C1 through the circuits of the timer. The timer motor does not run in this position because it is not being supplied with power. The contactor remains energized until the control relay CR is deenergized. The timer then restarts its cycle.

This timing method is one of several used as a protective circuit by the industry to prevent a heavy load from short cycling. This particular timer and timer relay circuit is one of the few methods that will reset itself automatically.

Compressor Circuits

The compressor circuit is shown in figure 3.49 along with timer and timer relay circuits. The equipment is using a part winding motor to operate the compressor. A part winding motor has two separate motors, which are energized at different times (although close together), built in one housing. The second motor or winding is energized a few seconds after the first motor or winding by a time delay relay. The motor is supplied by voltage through two contactors, C1 for the first motor or winding and C2 for the second.

The compressor circuits are equipped with circuit breakers CB1, CB2, and CB3 to protect the wiring and the motor. The power wiring is then split between the two contactors C1 and C2 that control the compressor. When the contacts of the contactor are closed, the compressor will operate.

The control circuit of the contactor is through the 208 volt-1 phase-60 cycle section of the diagram. This includes the CR contacts, the timer and timer relay circuits, the contactor coils, the time delay relay, and the crankcase heater. The crankcase heater operates when the normally closed contacts TR(1) are closed and the timer relay coil is deenergized.

The compressor contactors control the operation of the compressor through the timer and timer relay. Assume the timer motor has run its 4 minutes and 45 seconds on the "off" cycle and the timer is now in the positions A to A2 and B to B2. When the control relay contacts close, the timer starts again and the timer relay energizes, closing contacts TR(2) and TR(3) and keeping the timer relay energized until the control relay contacts open. After the timer has run for the 15 seconds, it switches to positions A to A1 and B to B1. This action energizes contactor coil C1 and the time delay relay TDR. The time delay relay closes about 2 seconds after C1 has energized and this energizes C2. Now the compressor is operating.

Condenser Fan Motor Circuits

The condenser fan motors OFM1, OFM2, and OFM3 are shown in the circuits of figure 3.50. The condenser fan motors are all controlled by the sepa-

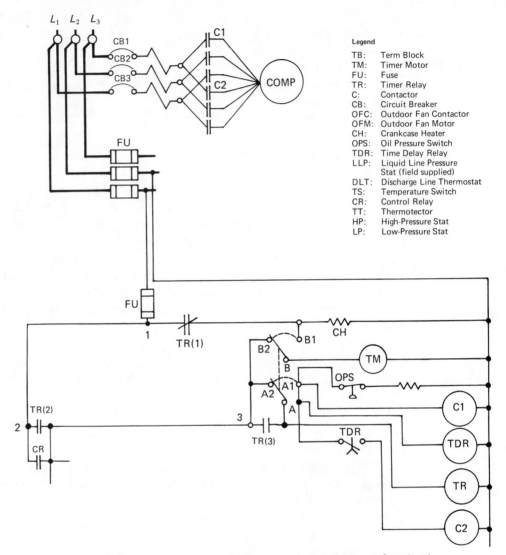

FIGURE 3.49.   Compressor circuits and timer and timer relay circuits.

rate contactors OFC1, OFC2, and OFC3. The first condenser fan motor contactor OFC1 is energized by the closing of the control relay contacts in circuit 3. When the coil of contactor OFC1 is energized, the condenser fan motor OFM1 is started. The additional two condenser fan motors operate by the temperature surrounding the unit. If the temperature is high enough to close temperature switch TS1, then contactor OFC2 is energized and starts fan motor OFM2. Fan motor OFM3 follows the same operation as fan motor OFM2.

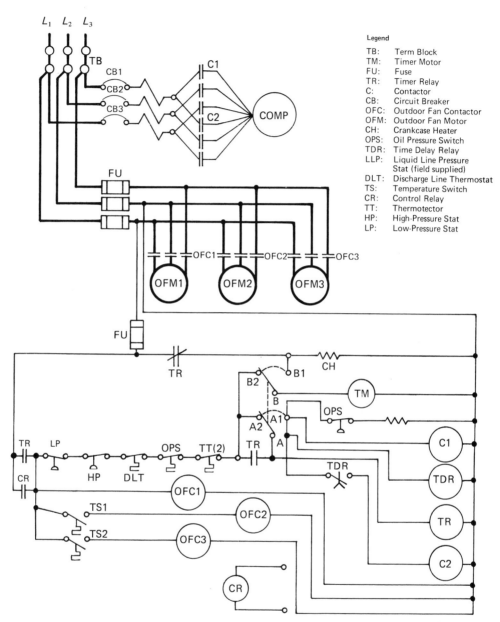

FIGURE 3.50. Condenser motor circuits, compressor circuits, and timer and timer relay circuits.

Circuit Safety Controls

The schematic diagram of the condensing unit in figure 3.50 shows all the safety devices. The safety devices discussed here will not include fuses and circuit breakers.

All the safety devices are added in one circuit except the heater of the oil pressure switch, which is connected between A1 and L_2. If any of the safety devices open, the compressor is stopped by the opening of the TR(3) contacts when the timer relay is deenergized because of the broken circuit. When this occurs the timer must complete its "off" cycle rotation of 4 minutes and 45 seconds.

The high pressurestat HP and low pressurestat LP maintain the pressure in the refrigeration system within safe limits and open if the pressure exceeds the safe limit. The discharge line thermostat DLT opens to protect the compressor if the temperature of the discharge line exceeds a set temperature. The thermotector TT is an internal motor-winding thermostat that protects the compressor motor from overheating by opening if the motor temperature exceeds a safe level.

The oil pressure switch OPS opens if the oil pressure of the compressor is too low for safe operation. The oil pressure switch that makes up the oil safety switch is controlled by a heater and a pressure switch shown in a circuit from A1 to L_2. If the oil pressure switch remains closed, the heater will heat up an element and open the switch of the OPS thermostat. If the oil pressure is high enough, the oil pressure switch OPS opens and cuts the heater out, allowing the unit to continue.

The schematic diagram covered in this chapter is not the most difficult diagram service personnel may encounter. However, it is a good example of what the service mechanic will come in contact with in the industry.

3.10   PICTORIAL DIAGRAMS

The **pictorial diagram**, also referred to as a label or line diagram, is intended to show the actual internal wiring of the unit. The pictorial diagram shows all the components of the control panel as a blueprint, including all the interconnecting wiring. It does not show the unit to scale, however. Components that are not shown in the control panel itself are shown outside the panel and labeled. The pictorial diagram is used basically to locate specific components or wires when troubleshooting from a schematic diagram. A typical pictorial diagram used in the industry is shown in figure 3.51.

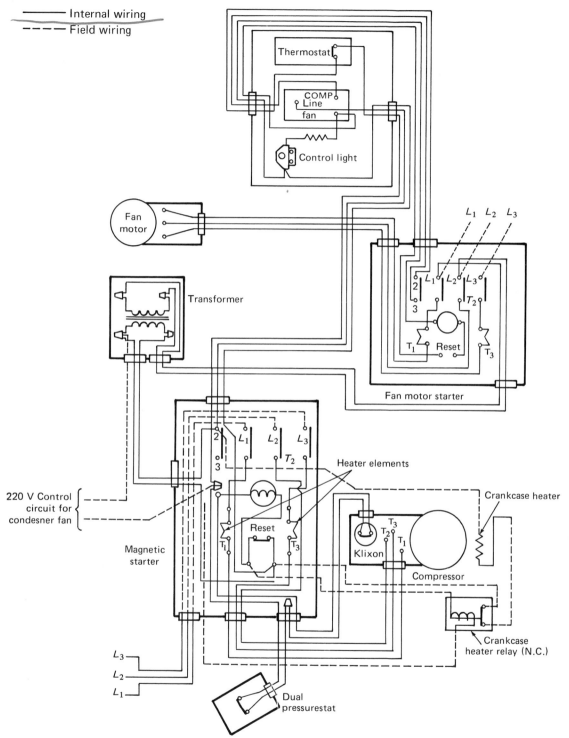

Internal wiring
Field wiring

FIGURE 3.51. A typical pictorial diagram used in the industry. Reproduced by permission of Carrier Corporation. © 1977 Carrier Corporation.

It is difficult to determine how a system operates by a pictorial diagram, and only an experienced mechanic can follow a complex pictorial diagram. Thus most air-conditioning personnel use the schematic diagram to find the cause of the problem. Then they use the pictorial diagram to locate the position of the component at fault. In cases where the wiring is simple, however, a pictorial diagram may be the only diagram furnished with the equipment.

The **factual diagram** consists of a pictorial diagram along with a schematic diagram. Many air conditioner manufacturers supply factual diagrams so that service personnel can locate the relay or component in the control panel.

3.11 INSTALLATION DIAGRAMS

The **installation diagram** is used to help the installation electrician to wire the unit properly. The diagram gives specific information about terminals, wire sizes, color coding, and breaker or fuse sizes. The diagram does not go into detail about equipment operation because the electrician has no need for this information. Figure 3.52 shows an installation diagram. The installation wiring diagram shows little internal wiring and is therefore almost useless to industry personnel.

SUMMARY

Loads are devices that use electricity to do useful work. Figure 3.53 gives a review of the symbols used for solenoids, motors, and heaters, the typical loads found in the industry. Most symbols have some type of letter designation to identify more clearly the component referred to.

Loads are controlled by relays and contactors, which share the same symbol and perform similar tasks. The major difference between relays and contactors is the amount of current each can carry. If a compressor is being operated by a device, you can assume that the device is a contactor. If a small fan motor is being operated by a device, you can assume the device is a relay. A relay is used for small loads and a contactor is used for large loads. Figure 3.54 reviews the symbols for some of these devices.

Relays and contactors are controlled by switches. Some of the switches used in the industry are manual, push button, thermostat, and pressure. Thermostats are made for two purposes: to operate a heating or a cooling system. The symbols for thermostats denote whether they are used for

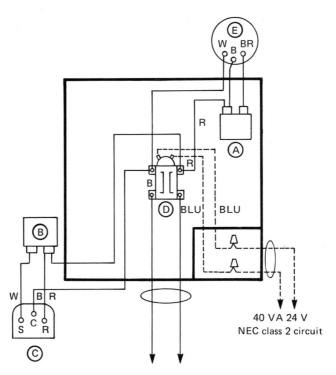

To power supply — 230 V (for-1A, -1C) 60 Hertz
To power supply — 230/208 V (for-2A, -2C) 60 Hertz
Single phase, 25 A Fusetron recommended not to
exceed 30 A

FIGURE 3.52. A typical installation diagram. Diagram courtesy of
The Singer Company, Climate Control Division.

FIGURE 3.53. Review of symbols used
for loads.

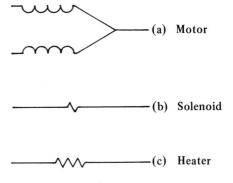

(a) Motor

(b) Solenoid

(c) Heater

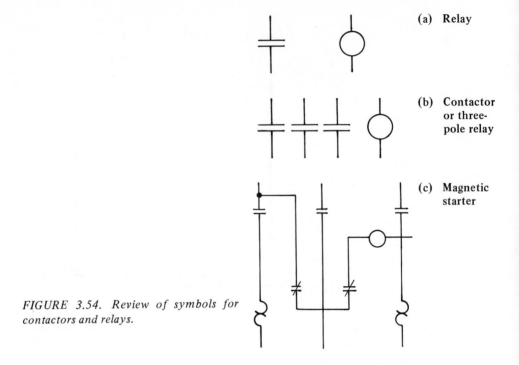

(a) Relay

(b) Contactor
     or three-
     pole relay

(c) Magnetic
     starter

FIGURE 3.54. Review of symbols for contactors and relays.

heating or cooling. Pressure switches are much the same as thermostats; their symbols also denote which way they open or close and under what condition. Pressure switches can be used for low or high pressure and are usually denoted by letter designations.

Protective devices are important in any system using motors to prevent damage to the motors or to larger components of the system. The most important type of safety devices are for motor protection. A fuse, magnetic overload, thermal overload line break, thermal overload pilot duty, or a thermal overload relay could be used. Many overloads are built directly into the larger components.

Transformers are devices that increase or decrease the incoming voltage to some desired voltage. Transformers are used in the industry mainly in control circuits.

Schematic diagrams tell air-conditioning, heating, or refrigeration personnel when and why a system works as it does. Schematic diagrams show the symbols for devices and the interconnecting wiring of a unit in a circuit-by-circuit arrangement. Schematic diagrams are used most frequently by service personnel to troubleshoot equipment and systems.

Pictorial diagrams show an exact layout of the control panel with the external components shown outside the panel and labeled. The pictorial diagram can be used as a troubleshooting diagram on a simple system, such as a window air conditioner. In most cases pictorial diagrams are used to find the placement of a component in the panel. Factual diagrams are a combination of the schematic and pictorial, with each shown separately.

Installation diagrams are used to help the installation electrician correctly connect the wiring to the unit.

QUESTIONS

1. What are the three types of electrical diagrams used in the heating, cooling, and refrigeration industry?

2. What is a load?

3. What is the major load of an air-conditioning system?

4. Identify the following symbols for loads:

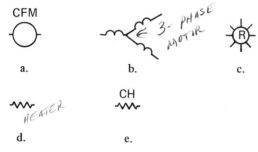

CFM

a.   b.   c.

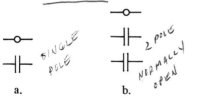

CH

d.   e.

5. What is the major difference between a contactor and a relay?

6. What do the terms "normally open" and "normally closed" refer to?

7. What is the difference between a magnetic starter and a contactor?

8. Identify the following symbols for relays and contactors:

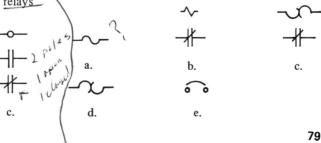

a.   b.   c.

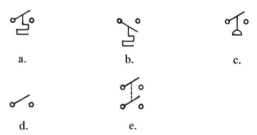

d.   e.

9. Why are heating and cooling thermostats different?

10. Identify the following symbols for switches:

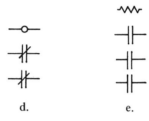

a.   b.   c.

d.   e.

11. What is the purpose of an overload?

12. What is a magnetic overload?

13. Identify the following symbols for safety devices:

a.   b.   c.

d.   e.

14. Draw the symbol for a transformer.

15. What are the requirements for an electric circuit?

16. What is a schematic diagram? What is its purpose?

17. What is a legend in a schematic diagram?

18. How is a pictorial diagram laid out?

19. What is a factual diagram?

20. What is the purpose of an installation diagram?

21. True or false. A solenoid is a device that opens or closes to control some element in a system. *EITHER*

22. Draw the symbol for a heater.

23. The purpose of the signal light in a system is to _____.

24. A contactor or relay is composed of _____.

25. True or false. Loads are controlled by switches, and switches are controlled by contactors or relays. *EITHER*

26. What do we mean when we say that a component is "energized"?

27. Name three basic switches used in the industry and draw their symbols.

28. What is the most important type of switch in a control system?

29. What is the purpose of a fuse in a system? What is the symbol for a fuse?

30. A bimetal element is _____.

31. True or false. A schematic diagram tells the service mechanic how to wire a system.

32. All schematic diagrams are broken down into _____.

4

Alternating Current, Power Distribution, and Voltage Systems

INTRODUCTION

There are two types of current used in the heating, cooling, and refrigeration industry today: **direct current** (DC) and **alternating current** (AC). Current is the flow of electrons. Direct current is an electron flow in only one direction. Alternating current is an alternating (back and forth) flow of electrons; that is, the electrons reverse their direction of flow at regular intervals. Direct current will not be discussed in this chapter because it is used in the industry only for special applications.

Almost all current produced by electric utilities is alternating current. In the rare instances that direct current is needed, it can be produced by a direct current generator or a rectifier. It is usually produced by the consumer. Direct current has limited use in the industry; it is used mostly for refrigeration transportation equipment, electronic air cleaners, and electronic control components. *SOLID STATE*

Alternating current is used in most heating, cooling, and refrigeration equipment. Alternating current equipment is cheaper, more rugged, and more trouble free than direct current equipment. In addition, alternating current is easier to produce than direct current.

Because of the popularity and wide use of alternating current, it is important for industry personnel to be familiar with its theory. In addition, personnel should be familiar with the way power is distributed by electric utilities and the many types of voltage-current systems available.

We begin our study with a discussion of some of the basic ideas of alternating current.

4.1 BASIC CONCEPTS OF ALTERNATING CURRENT

Alternating current, or AC, is an electron flow that alternates, flowing in one direction and then in the opposite direction at regular intervals. It is produced by cutting a magnetic field with a conductor.

The **sine wave** is often used as a graphical representation of alternating current. Figure 4.1 shows a sine wave, the graph of alternating current. The letter X represents a conductor in a particular position as it is rotated in the magnetic field. At point *A*, X is at a potential of 0. However, the potential (voltage) increases when the conductor is rotated through the magnetic field until it reaches point *B*, where the potential peaks. When the conductor (X) is rotated from *B* to *C*, the potential decreases until at point *C* the potential is again 0. The direction of flow from point *A* to point *C* is termed positive. The direction is reversed on the lower half of the sine wave (point *C* back to point *A*) and is termed negative. From point *C* the potential increases until it reaches point *D* and peaks. The potential decreases from point *D* until the conductor reaches point *A*, where the potential is again 0. The curve shows

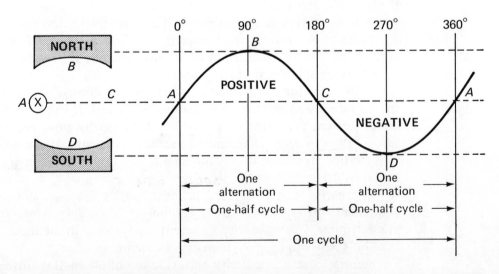

FIGURE 4.1. Sine wave of alternating current.

the conductor from point A through a complete clockwise revolution back to point A.

While the conductor is rotating from point A to point C, one positive alternation occurs. That is, the curve starts from 0, reaches the peak at point B, and returns to 0 at point C. The next alternation is negative as X goes from point C back to point A in the same manner as the first alternation.

Cycles and Frequency

When the conductor rotates through one complete revolution, it has generated two alternations, or flow reversals. Two alternations (changes in direction) equal one cycle, as shown in figure 4.1. One cycle occurs when the **rotor**, or conductor, cuts the magnetic field of a north pole and a south pole. SAME The **frequency** of alternating current is the number of complete cycles that occur in a second. The frequency is known as **hertz** (Hz), but many times it is referred to as **cycles**. In almost all locations in the United States, the common frequency is 60 hertz. In some isolated cases in the United States and in some foreign countries, 25 hertz is the frequency used. The disadvantage to using 25 hertz is that the reversals can be detected by the human eye in light fixtures. The frequency of 60 hertz is considered standard frequency, but 50 hertz is also used in some cases.

Effective Voltage

Because alternating current starts at 0, reaches a peak, and then returns to 0, there is always a variation in voltage and an effective value has to be determined. Alternating current reaches a peak at 90 electrical degrees. This is known as **peak voltage**. The **effective voltage** of an alternating current circuit is 0.707 times its highest or peak voltage. This value is determined with respect to direct current so that the effective voltage is equal to one direct current volt. Figure 4.2 shows a comparison of effective voltage versus peak voltage in an alternating current circuit. Alternating current amperage is 0.707 times the peak amperage. All electric meters are calibrated to read effective voltage and amperage.

Voltage-Current Systems

Alternating current is available to the consumer at several different voltages and with different current characteristics. The four basic voltage-current

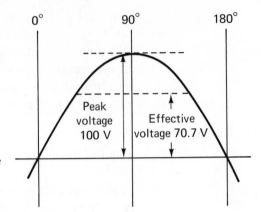

FIGURE 4.2. *Effective voltage versus peak voltage.*

characteristics available are 230 volt–single phase–60 hertz, 230 volt–three phase–60 hertz, 208 volt–three phase–60 hertz, and 460 volt–three phase–60 hertz. These designations are often abbreviated; for example, 230 volt–single phase–60 hertz is abbreviated as 230 V–1ϕ–60 Hz. The V is the abbreviation for volts; ϕ (Greek letter phi) is the symbol for phase; and Hz is the abbreviation for hertz (which means the same as cycles).

Phase

The **phase** of an AC circuit is the number of currents alternating at different time intervals in the circuit. Single-phase current would have only a single current, while three-phase current would have three.

Inductance and Reactance

It is a common assumption that when the voltage is strongest, the amperage is also strongest. However, this is not always the case. The amperage of a circuit determines the strength of the magnetic field in the circuit. If the current increases, so will the magnetism. If the current decreases, so will the magnetism. The fluctuation of the magnetic strengths in an AC circuit, and in conductors cutting through more than one magnetic field, *induces* (causes) a voltage that counteracts the original voltage. This effect is called **inductance**.

The inductance in an AC circuit—that is, the effect of the magnetic fields—produces an out-of-phase condition between the voltage and amperage. The induction of the original voltage produces a second voltage due to the magnetic fields collapsing. This effect causes the voltage to *lead* the amperage, as shown in figure 4.3.

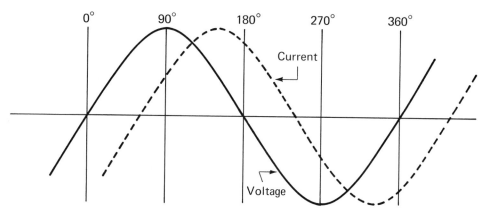

FIGURE 4.3. *Sine wave of voltage and current when they are out of phase; the voltage leads the current in an AC circuit due to the effect of inductance.*

Resistance in a direct current circuit is the only factor that affects the current flow. AC circuits are affected by resistance, but they are also affected by **reactance**. Reactance is the resistance that alternating current encounters when it changes flow. There are two types of reactance in AC circuits: inductive reactance and capacitive reactance. **Inductive reactance** is the opposition to the change in flow of alternating current, which produces an out-of-phase condition between voltage and amperage, as shown in figure 4.3. **Capacitive reactance** is caused in AC circuits by using capacitors. When a capacitor is put in an AC circuit, it resists the change in voltage, causing the amperage to lead the voltage, as shown in figure 4.4. The sum of the resistance and reactance in an AC circuit is called the impedance. *TEST*

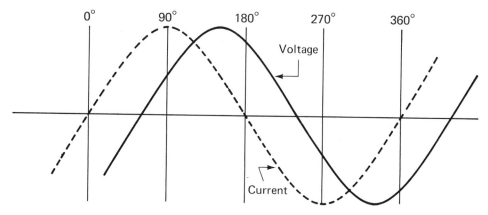

FIGURE 4.4. *Sine wave of voltage and current when they are out of phase; the current leads the voltage in an AC circuit due to capacitive reactance.*

85

Power

As we have said, the voltage and current in an AC circuit are out of phase with each other. When voltage and current are out of phase, they are not working together. Thus the power (wattage) of the circuit must be calculated by using a voltmeter and an ammeter. This calculation gives the **apparent wattage**. A watt meter would measure the true power (wattage). The ratio between the true power and the apparent power is called the **power factor** and is usually expressed as a percentage. The following equation expresses this:

$$PF = \frac{\text{true power}}{\text{apparent power}}$$

4.2 POWER DISTRIBUTION

Direct current was used in the beginning to supply consumers with their electrical needs, but it had many disadvantages. Transmission for a long distance without using generating stations to boost the direct current was impossible. The inability to raise and lower DC voltages and the necessity to use large transmission equipment were other problems. Alternating current can be transmitted with much less worry than direct current and has become the ideal power to supply to consumers because of its flexibility. Almost all equipment utilized in the industry incorporates alternating current as its main source of power.

Electric power is generated by rotating turbines through the use of gas, oil, coal, hydropower, or atomic energy. The rotating turbines have the effect of a rotating conductor in a magnetic field. The electricity is generated inside the plant and transferred outside the plant, where it is boosted to a large, easily transmitted voltage. This is frequently as high as 220,000 volts. The alternating current is then transmitted to a substation, where its voltage is reduced to around 4,800 volts with the use of a step-down transformer. From the substation the power is distributed to transformers that step down the voltage to a usable voltage. Or the power is supplied directly to consumers who use their own transformers to reduce the voltage. Figure 4.5 shows a diagram of electric power transmission from the generating plant to the consumer.

In the following sections we will discuss the four common voltage systems available to consumers.

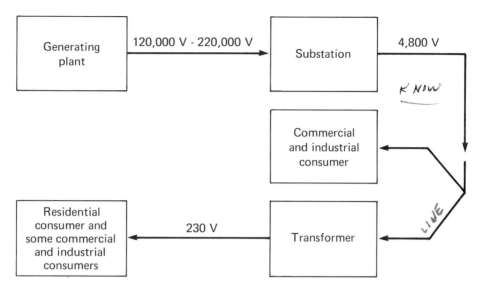

FIGURE 4.5.   Layout of power distribution.

4.3   230 VOLT–SINGLE PHASE–60 HERTZ SYSTEMS

Single-phase alternating current exists in almost all residences. Almost all small heating, cooling, and refrigeration equipment can be purchased for use on single phase. Any domestic appliance that operates on 110 volts is single-phase equipment. It is important to understand the makeup of the 230 V–1ϕ–60 Hz system because of its popularity in small commercial buildings and residences. Air-conditioning personnel must be able to determine the type of voltage system available to ensure proper installation and operation of the equipment.

In some older structures it is still possible to find a single-phase–two-wire system. This voltage system was used when few electric appliances were available and there was no need for 230-volt systems. The system uses two wires, one as a hot wire and one as a neutral. **Hot wires** are conductors that actually supply voltage to an appliance. The neutral is a wire connected to the ground. In the voltage system shown in figure 4.6, 110 volts is fed through L_1 but must be connected to a neutral to make a complete circuit.

The most common voltage system found today is the 230 V–1ϕ–60 Hz system. This voltage system consists of three wires, two hot wires and one ground. A schematic for the 230-volt system is shown in figure 4.7. The figure shows the system being used as a switch to break down the power being fed to a piece of equipment, such as a walk-in cooler, an electric fur-

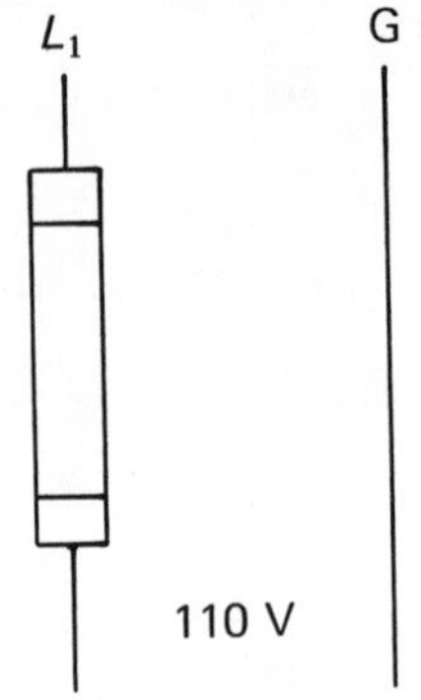

FIGURE 4.6. *Schematic for a 110 volt-single phase–60 hertz system using two wires.*

nace, or an air-conditioning unit. In this system either of the hot legs L_1 or L_2, when connected to the neutral, will supply 110 volts. Connecting L_1 and L_2 directly to a load will supply 230 volts.

The electric utility uses a transformer to produce the 230 V–1ϕ–60 Hz system. The transformer hookup of this system is shown in figure 4.8. Pay close attention to the secondary side of the transformer. A transformer takes voltage at one value and by induction changes it to another value. Figure 4.8 shows a transformer with a 4800-volt primary winding and a 230-volt secondary winding. The primary winding, or input, is connected to the initial voltage, which is 4800 volts in figure 4.8. The secondary winding produces the output, or new voltage, which is 230 volts in figure 4.8.

All equipment can be operated at plus or minus 10% of the rated voltage. For example, a piece of equipment rated at 230 volts could operate on a minimum voltage of 207 volts and a maximum voltage of 253 volts. The electric utility maintains a plus or minus 10% of the correct supply voltage.

FIGURE 4.7. *Schematic for a 230 volt-single phase–60 hertz system using three wires.*

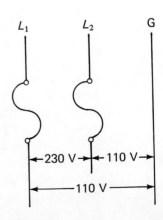

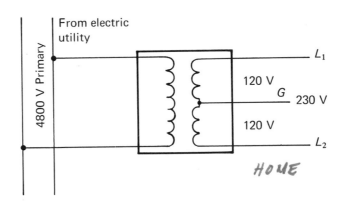

FIGURE 4.8.   Transformer hookup of a 230 V–1φ–60 Hz system.

Air-conditioning equipment has a tendency to operate more satisfactorily on a maximum voltage than on the minimum voltage.

4.4 THREE-PHASE VOLTAGE SYSTEMS

Three-phase alternating current is common in most commercial and industrial structures. Three-phase electrical services supply three hot legs of power with one ground to the distribution equipment and then on to the equipment. Three-phase power supplies are more versatile than single-phase supplies because of the different voltage systems that are available. Thus it is essential to understand the uses and advantages of three-phase voltage systems.

Three-phase electric power supplies have many advantages over the single-phase power supply when used in commercial and industrial applications. In most cases residences do not use enough electric energy to warrant a three-phase power supply. The electric power consumer can buy three-phase electric power cheaper than single-phase. Three-phase electric motors require no special starting apparatus, eliminating one trouble spot in the building and servicing of motors. Three-phase power offers better starting and running characteristics for motors than does single-phase. Many of the large electric motors used in the industry are available for three-phase only, preventing many structures from using single-phase power supplies. Thus almost all commercial and industrial structures are supplied with three-phase power.

The only disadvantage to three-phase systems is the higher cost of electric panels and distribution equipment.

There are basically two types of three-phase voltage systems used in commercial and industrial wiring systems: the **delta transformer** hookup, which will supply 230 V–3φ–60 Hz, and the **Y transformer** hookup, which will supply 208 V–3φ–60 Hz or 460 V–3φ–60 Hz.

4.5 230 VOLT–THREE PHASE–60 HERTZ SYSTEMS – DELTA

The 230 V–3ϕ–60 Hz system is used basically in structures that require a large supply to motors and other three-phase equipment. The delta system is usually supplied to a structure with four wires, which include three hot legs and a ground, but in some rare cases it is supplied with three hot wires only. Figure 4.9 shows the wiring layout of a 230 V–3ϕ–60 Hz system. This system is unique in that it contains a high leg, which in figure 4.9 is L_4. Connecting L_1 to neutral, or ground, provides a range of 180 volts to 208 volts. Connecting L_1 to L_2, L_2 to L_3, or L_1 to L_3 provides 230 volts. Connecting L_2 to G or L_3 to G provides 120 volts. If one hot leg of a three-phase system is lost, only single-phase voltage is supplied and a single-phase condition exists. This condition can easily damage any three-phase equipment.

The transformer secondary hookup of a three-phase delta system is shown in figure 4.10. The delta system takes its name from the Greek letter delta Δ, which resembles the shape of the hookup, as can be seen in figure 4.10. The high leg of the voltage is due to the fact that the transformer winding between L_1 and G is longer than the windings between L_2 and G and between L_3 and G.

The 230 V–3ϕ–60 Hz, or delta, voltage system is used primarily on systems that have many three-phase circuits, of 230-volt circuits, with few 115-volt circuits. The system consists of two available 120-volt legs when L_2 and L_3 are connected to a circuit with the ground. If the high leg, or L_1, is connected in a circuit with the ground, it delivers between 180 volts and 208 volts. This will damage any 115-volt load or appliance.

Detection of the 230 V–3ϕ–60 Hz system can be accomplished simply and easily by testing across any two hot legs with a voltmeter. If 230 volts

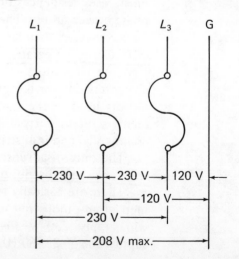

FIGURE 4.9. Schematic for a 230 volt-three phase–60 hertz system using four wires.

DELTA

LOT OF MOTORS

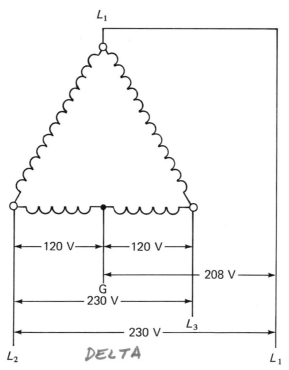

L_1

120 V — 120 V

208 V

G
230 V

230 V L_3

L_2 DELTA L_1

FIGURE 4.10. Schematic for the transformer hookup of a 230 V–3ϕ–60 Hz system showing the delta transformer secondary hookup.

are read, the system is a delta system. The high leg is usually located on the right side of the electric distribution equipment. Another method of detection is to check between all three hot legs and the ground with a voltmeter. If any one of the three hot legs between the ground reads between 180 volts and 208 volts, it is a delta system.

DELTA Y TRANSFORMER

4.6 208 VOLT-THREE PHASE-60 HERTZ SYSTEMS Y

The 208 V–3ϕ–60 Hz voltage system is common in structures that require a large number of 110-volt circuits, such as in schools, hospitals, and office buildings. The system offers the versatility of using three-phase alternating current and the possibility of supplying many 110-volt circuits for lights, appliances, and other 110-volt equipment. Voltage ratings of some equipment are such that the equipment must be used with the right voltage system for proper operation and selection. Therefore detection of this system is important because of the lower-voltage requirements.

This system is supplied by four wires, one ground and three hot legs, as shown in schematic form in figure 4.11. The 208-volt system is different from the 230-volt system in that it contains no high leg. In figure 4.11, L_1, L_2, and L_3 are the hot legs and G represents the ground. The voltage available between any two hot legs (L_1, L_2, or L_3) is 208 volts. Connections between any hot leg (L_1, L_2, or L_3) and the ground (G) provides 110 volts. As can be seen from the schematic, there are three available 110-volt power legs. This allows more 110-volt circuits than the 230-volt delta system.

The transformer secondary hookup of a 208-volt–three-phase Y system is shown in figure 4.12. The Y system takes its name from the letter Y, which resembles the shape of the hookup, as can be seen in figure 4.12. All three legs of the transformer windings are equal in winding length, as shown in figure 4.12, but in the delta system they are not. The Y transformer connection is also used on several higher-voltage systems.

Detection of the 208 V–3ϕ–60 Hz Y system is easy with the use of a voltmeter. If 208 volts are read across any two of the hot legs, or if 110 volts are read between all three hot legs and the ground, the system is a 208-volt Y system. The 208-volt Y system is a balanced system and contains no high leg that must be identified. But care should be taken to identify the ground when connections are being made.

4.7 HIGHER-VOLTAGE SYSTEMS — FACTORY

Higher-voltage systems are becoming increasingly popular because of their many advantages. The higher-voltage systems are used mostly on industrial

FIGURE 4.11. Schematic for a 208 volt-three phase-60 hertz system using four wires.

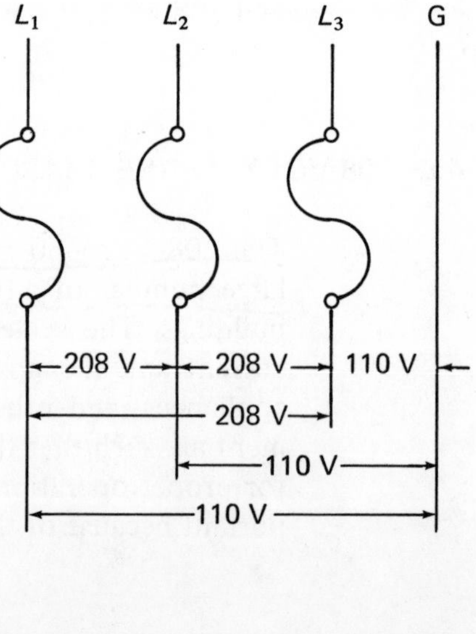

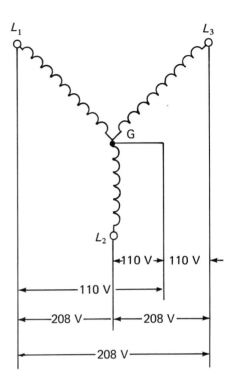

FIGURE 4.12.   Schematic for the transformer hookup of a 208 V–3φ–60 Hz system showing the Y transformer secondary hookup.

structures, but in some cases they are used on commercial systems. It is necessary to understand and to be able to detect the higher-voltage system to ensure the safety of personnel and to prevent damage to the equipment.

There are several high-voltage systems available. There is a 240/480-volt–single-phase system, a 240/416-volt–three-phase system, and a 277/480-volt–three-phase system. These systems can be identified with a voltmeter. For example, in the 277/480-volt system, the 277-volt reading is obtained between any two hot legs. The different systems are available with Y and delta hookups. The Y 277/480-volt system will be discussed in this section because of its popularity, but other high-voltage systems are available.

Advantages

There are many advantages to using higher-voltage systems. First, all wiring is rated at 600 volts. Second, there is little difference in the switches, relays, and other electric panels used in 208-volt and 480-volt systems. Motors may be wound differently for higher voltages, but they cost little more. The service equipment and wiring may be smaller for a 480-volt system than for a 208-volt system. This might save the consumer a great deal of money.

Disadvantages

Disadvantages to using the higher-voltage systems stem mainly from problems that would be brought about by trying to implement a common high-voltage system in this country. If the consumer needs single-phase circuits with both 110 volts and 230 volts, additional transformers must be purchased. Single-phase equipment is far more common than three-phase. Equipment manufacturers would like to work toward some common voltages so that they could produce fewer types of equipment. Now they must manufacture many motors with different voltage ratings to meet consumers' needs.

The 277/480-Volt System

Figure 4.13 shows a schematic layout of the 277/480-volt Y system. The system is primarily used in industry, but occasionally there are commercial applications. The system has no means of supplying 110 volts or 230 volts at single phase without the use of a separate step-down transformer.

Between any hot leg (L_1, L_2, or L_3) and ground (G), 277-volt circuits are obtained. The 277-volt circuit is commonly used in commercial and industrial lighting systems to operate fluorescent lights. Between any two hot legs (L_1 and L_2, L_2 and L_3, or L_1 and L_3), 480 volts can be obtained. If three-phase is desirable, it can be obtained by connections to all three hot legs. Figure 4.14 shows the completely balanced transformer layout of a 277/480 V–3ϕ–60 Hz Y system.

Detection of the 480 V–3ϕ–60 Hz system is easy with the use of a volt-

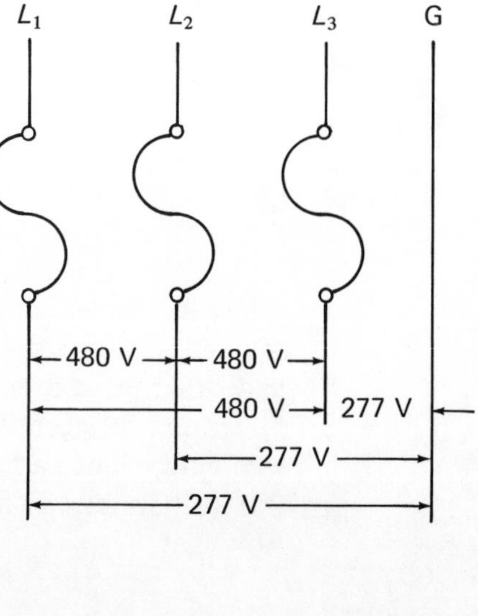

FIGURE 4.13.   Schematic for a 277/480 V–3ϕ–60 Hz Y system.

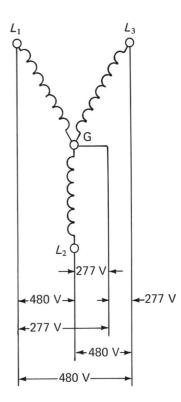

FIGURE 4.14. *Schematic for the transformer hookup of a 277/480 V–3φ–60 Hz Y system.*

meter. If 480 volts are read across any two of the hot legs, or if 277 volts are read between the hot legs and the ground, the system is a 277/480-volt Y system. The 480-volt Y system is a balanced system and contains no high leg that must be identified. But care should be taken to identify the ground when connections are being made.

SUMMARY

Alternating current is used in almost all heating, cooling, and refrigeration equipment because its properties give it greater flexibility than direct current. It is easier to transmit over a long distance and no expensive transmission equipment is required. It can be supplied at almost any voltage that a consumer wants.

Alternating current changes direction twice per cycle. The number of cycles per second is the frequency. Standard frequency is 60 hertz. The voltage leads the current in an AC circuit due to inductive reactance. The current leads the voltage due to capacitive reactance. The sum of the resistance and reactance in an AC circuit is called the impedance.

Voltage and current are out of phase in an AC circuit. Consequently, the power of the circuit must be calculated by using the effective wattage. The ratio between the true power and the effective power in an AC circuit is called the power factor.

An electric utility company can supply single-phase or three-phase current to structures at voltages from 208 volts to 460 volts. The generating plant supplies a high voltage that is stepped down so that the consumer can use it. The common voltage characteristics are 230 V–1ϕ–60 Hz, 230 V–3ϕ–60 Hz, 208 V–3ϕ–60 Hz, and 460 V–3ϕ–60 Hz.

The 230 V–3ϕ–60 Hz, or delta, system is common throughout commercial and industrial structures. It adds efficiency to systems that have a large number of three-phase and 230-volt circuits. The system supplies three hot legs and one ground, which deliver 230 volts between any two of the three hot legs. This system has a high leg. Caution should be taken not to connect any 110-volt circuit to the high leg. Between ground and the other two hot legs, 110 volts can be obtained. Detection of the delta system is essential to prevent damage to 110-volt equipment and to ensure proper equipment selection.

The 208-volt Y system is a well-balanced three-phase system that is commonly used on structures that need many 110-volt circuits as well as three-phase or 208-volt circuits. The system does not operate motors as efficiently as the 230-volt delta system, but it is more adaptable to the balancing of each leg of a service. Much equipment is made for either 208-volt or 230-volt systems and some can be used on either system.

High-voltage systems are becoming increasingly common. In a high-voltage system smaller wire can be used, making installation of equipment less expensive. While there are several high-voltage systems in use, the 277/480-volt Y system is most common. There is a great deal of heating, cooling, and refrigeration equipment rated at higher voltages. Air-conditioning personnel should be able to identify the higher-voltage systems for safety and proper installation.

QUESTIONS

1. What is alternating current?

2. What is the advantage of alternating current over direct current?

3. The common frequency of alternating current used in the United States is _____.

4. What is reactance?

5. What is impedance?

6. Give a brief description of the transmission of alternating current from the generating plant to the consumer.

7. Name the four voltage-current characteristics that are commonly used today.

8. The phase of an alternating current is _____.

9. True or false. Inductance in an AC circuit is an effect due to the magnetic fields caused by the current flow.

10. What is the common electrical service to a residence?

11. What is the voltage range on which alternating current equipment can be operated?

12. How can a 230 V–1ϕ–60 Hz electric system be detected?

13. What is the difference between single-phase and three-phase alternating current?

14. What is the voltage of a system with a Y transformer hookup?

15. What is the voltage of a system with a delta transformer hookup?

16. What are the advantages and disadvantages of a delta electric system?

17. How can the Y and delta systems be detected?

18. What is the advantage of the 460-volt Y system over the 208-volt Y system?

19. True or false. Single-phase alternating current is common in most commercial and industrial structures.

20. Sketch the delta and Y transformer hookup arrangements.

21. What are the advantages of the 208 V–3ϕ–60 Hz system? In which types of structures is this system primarily used?

5

Installation of Heating, Cooling, and Refrigeration Systems

INTRODUCTION

The proper installation of heating, cooling, and refrigeration equipment is just as important as any other phase of the industry. Installation covers a broad range of subjects, but one of the most important is the electric circuit servicing the equipment and its size. Thus industry personnel should be familiar with the structure's circuitry and circuit components.

Once the electric utility delivers the power to the structure, the consumer must bring it inside. There are several kinds of electric panels used in residences and many types used in commercial and industrial structures to accomplish this. Hence it is important for service and installation technicians to understand how electric power is distributed within the structure.

In this chapter we will discuss several of the components of the electric circuit servicing the heating, cooling, and refrigeration equipment. We will also discuss several types of electric panels that personnel may encounter on the job.

5.1 SIZING WIRE

Manufacturers usually list in the installation instructions the correct wire and fuse size. But in many cases the person responsible for the installation must calculate the wire and fuse size. The National Electrical Code governs the types and sizes of wire that can be used for a particular application and a certain amperage. The correct wire and fuse size is important to the life and efficiency of any equipment. Hence the installation mechanic should know how to determine the size to use.

Copper is the most popular conductor in the industry. However, aluminum is used in some cases because of its low cost. Copper wire is a good conductor of electricity and has many other characteristics that make it the most popular conductor used. Copper wire bends easily, has good mechanical strength, has the ability to resist corrosion, and can be easily joined together. Aluminum conductors, on the other hand, do not have all the good characteristics of copper. Aluminum conducts electricity well enough, but problems arise because aluminum corrodes easily. Thus aluminum wire connections have a tendency to become loose after they have corroded.

Wire Size

Standard wire size is defined in the United States by the American Wire Gauge (AWG). The American Wire Gauge lists the largest wire, 0000 (4/0), down to number 50, which is the smallest wire. In the industry, wire sizes from number 20 to number 4/0 are the most common. The most popular sizes are from number 16 to number 4.

In some cases wire larger than 4/0 is needed. The circular mil system is used for this purpose. Circular mil sizing runs from 250 MCM (MCM is the abbreviation for 1000 circular mils), which is about ½ inch in diameter, to 750 MCM, which is about 1 inch in diameter. Circular mil sizing does exist in larger sizes. Figure 5.1 is a table that gives the data on round copper wire.

Factors to Consider in Wiring — 4

The type of insulation surrounding the conductor usually determines its application and the amperage it can be used for. Insulation of different grades is used for different purposes. For example, insulation can be heat resistant, moisture resistant, heat and moisture resistant, or oil resistant. Figure 5.2 shows a table from the National Electrical Code (NEC) listing conductor application and insulations.

There are several factors that should be considered when sizing circuit conductors. These factors are voltage drop, insulation type, enclosure, and safety. The voltage drop of a circuit, which takes into account the distance the conductors must be fed, must be calculated by the installation technician.

American Wire Gauge (A.W.G.) Working Table (U.S. Bureau of Standards)*

| Gauge No. A.W.G. | Diameter in Mils | Cross Section | | Ohms per 1000 Ft. at 25 Deg. C. (77 Deg. F.) | Lb. per 1000 Ft. |
		Circular Mils	Square Inches		
0000	460	212000	.166	.0500	641
000	410	168000	.132	.0630	508
00	365	133000	.105	.0795	403
0	325	106000	.0829	.100	319
1	289	83700	.0657	.126	253
2	258	66400	.0521	.159	201
3	229	52600	.0413	.201	159
4	204	41700	.0328	.253	126
5	182	33100	.0260	.320	100
6	162	26300	.0206	.403	79.5
7	144	20800	.0164	.508	63.0
8	128	16500	.0130	.641	50.0
9	114	13100	.0103	.808	39.6
10	102	10400	.00815	1.02	31.4
11	91	8230	.00647	1.28	24.9
12	81	6530	.00513	1.62	19.8
13	72	5180	.00407	2.04	15.7
14	64	4110	.00323	2.58	12.4
15	57	3260	.00256	3.25	9.86
16	51	2580	.00203	4.09	7.82
17	45	2050	.00161	5.16	6.20
18	40	1620	.00128	6.51	4.92
19	36	1290	.00101	8.21	3.90
20	32	1020	.000802	10.4	3.09
21	28.5	810	.000636	13.1	2.45
22	25.3	642	.000505	16.5	1.94
23	22.6	509	.000400	20.8	1.54
24	20.1	404	.000317	26.2	1.22
25	17.9	320	.000252	33.0	0.970
26	15.9	254	.000200	41.6	0.769
27	14.2	202	.000158	52.5	0.610
28	12.6	160	.000126	66.2	0.484
29	11.3	127	.0000995	83.5	0.384
30	10.0	101.0	.0000789	105	0.304
31	8.9	79.7	.0000626	133	0.241
32	8.0	63.2	.0000496	167	0.191
33	7.1	50.1	.0000394	211	0.152
34	6.3	39.8	.0000312	266	0.120
35	5.6	31.5	.0000248	336	0.0954
36	5.0	25.0	.0000196	423	0.0757
37	4.5	19.8	.0000156	533	0.0600
38	4.0	15.7	.0000123	673	0.0476
39	3.5	12.5	.0000098	848	0.0377
40	3.1	9.9	.0000078	1070	0.0299

FIGURE 5.1. Data on round copper wire. Chart courtesy of Continental Wire & Cable, Div. of the Anaconda Co.

The insulation type, the enclosure, and safety can be determined by using the tables of the *National Electrical Code.* There are also wire-sizing tables in the *National Electrical Code* that give the allowable amperage for both aluminum and copper conductors. The *National Electrical Code* is considered a guide to safe wiring procedures. The code does not ensure that all systems following its procedures will be good systems, however. That is the responsibility of the designer.

Voltage Drop

Voltage drop in a conductor is of prime importance when sizing wire. Any voltage that drops between the supply and the equipment is lost to the equipment. If the voltage drop is large enough, it will seriously affect the operation of the equipment. But even a small voltage drop is detrimental to the equipment. The voltage drop can be easily measured when the equipment is operating by reading the voltage at the supply and subtracting from that the voltage read at the equipment. If we read 240 volts at the supply and 210 volts at the equipment when it is operating, then there is a voltage drop of 30 volts in the circuit. The allowable voltage drop of most equipment is 2%. - 3% IDEAL   < 10% - OK

equipment ± 10% will run

Wire-Sizing Charts

Figure 5.3 shows the table from the *National Electrical Code* for the allowable ampacities (amperages) of a conductor for copper and aluminum wire. The wire-sizing charts from the NEC are usually accurate for sizing electric

---

* The fundamental resistivity used in calculating the table is the annealed-copper standard, via., 0.15328 ohm (meter, gram) at 20 deg C. The temperature coefficient for this particular resistivity at 20 deg. C. = 0.00393, or at 0 deg. C. = 0.00427. However, the temperature coefficient is proportional to the conductivity and hence the change of resistivity per degree C, is a constant, 0.000597, ohm (meter, gram). The "constant mass" temperature coefficient of any sample is

$$aXt = \frac{0.000597 \; + \; 0.000005}{\text{resistivity in ohms (meter, gram) at t degrees C.}}$$

The density is 8.89 grams per cubic centimeter.

The values given in the table are only for annealed copper of the standard resistivity. The user of the table must apply the proper correction for copper or any other resistivity. Hard, drawn copper. may be taken as about 2.7 per cent higher resistivity than annealed copper.

Ohms per mile, or pounds per mile, may be obtained by multiplying the respective values above by 5.28.

† Resistance at T degrees above or below 25 deg. C. = R (at 25 degrees) (1 + 0.00385 [T-25]).

The values given for ohms per 1000 feet and pounds per 1000 feet are 2 per cent greater than for a solid rod of cross section equal to the total cross section of the wires of the cable.

Resistivity of pure copper at 20 deg C. = 0.15328 ohms per meter, gram.

Source: Continental Wire and Cable.

Trade Name	Type Letter	Max. Operating Temp.	Application Provisions	Insulation	AWG or MCM	Thickness of Insulation (Mils)	Outer Covering
Heat-Resistant Rubber	RH	75°C 167°F	Dry locations.	Heat-Resistant Rubber	**14-12 10 8-2 1-4/0 213-500 501-1000 1001-2000	30 45 60 80 95 110 125	*Moisture-resistant, flame-retardant, non-metallic covering
Heat-Resistant Rubber	RHH	90°C 194°F	Dry locations.	Heat-Resistant Rubber	14-10 8-2 1-4/0 213-500 501-1000 1001-2000	45 60 80 95 110 125	Moisture-resistant, flame-retardant, non-metallic covering
Moisture and Heat-Resistant Rubber	RHW	75°C 167°F	Dry and wet locations. For over 2000 volts insulation shall be ozone-resistant.	Moisture and Heat Resistant Rubber	14-10 8-2 1-4/0 213-500 501-1000 1001-2000	45 60 80 95 110 125	*Moisture-resistant, flame-retardant, non-metallic covering
Heat-Resistant Latex Rubber	RUH	75°C 167°F	Dry locations.	90% Un-milled, Grainless Rubber	14-10 8-2	18 25	Moisture-resistant, flame-retardant, non-metallic covering
Moisture-Resistant Latex Rubber	RUW	60°C 140°F	Dry and wet locations.	90% Un-milled, Grainless Rubber	14-10 8-2	18 25	Moisture-resistant, flame-retardant, non-metallic covering
Thermoplastic	T	60°C 140°F	Dry locations.	Flame-Retardant, Thermoplastic Compound	14-10 8 6-2 1-4/0 213-500 501-1000 1001-2000	30 45 60 80 95 110 125	None
Moisture-Resistant Thermoplastic	TW	60°C 140°F	Dry and wet locations.	Flame-Retardant Moisture-Resistant Thermoplastic	14-10 8 6-2 1-4/0 213-500 501-1000 1001-2000	30 45 60 80 95 110 125	None
Heat-Resistant Thermoplastic	THHN	90°C 194°F	Dry locations.	Flame-Retardant Heat-Resistant	14-12 10 8-6 4-2	15 20 30 40	Nylon jacket

Table of conductor application and insulations. The table is presented in landscape orientation.

Insulation	Trade Name	Max Operating Temp	Application Provisions	Insulation Type	Size (AWG/kcmil)	Thickness (mils)	Outer Covering*
Moisture- and Heat-Resistant Thermoplastic	THW	75°C / 167°F 90°C / 194°F	Dry and wet locations. Special applications *within* electric discharge lighting equipment. Limited to 1000 open-circuit volts or less. (Size 14-8 only as permitted in Section 410-26).	Thermoplastic	1-4/0 250-500 501-1000	50 60 70	None
				Flame-Retardant, Moisture- and Heat-Resistant Thermoplastic	14-10 8-2 1-4/0 213-500 501-1000 1001-2000	45 60 80 95 110 125	
Moisture- and Heat-Resistant Thermoplastic	THWN	75°C / 167°F	Dry and wet locations.	Flame-Retardant, Moisture- and Heat-Resistant Thermoplastic	14-12 10 8-6 4-2 1-4/0 250-500 501-1000	15 20 30 40 50 60 70	Nylon jacket
Moisture- and Heat-Resistant Cross-Linked Synthetic Polymer	XHHW	90°C / 194°F 75°C / 167°F	Dry locations. Wet locations.	Flame-Retardant Cross-Linked Synthetic Polymer	14-10 8-2 1-4/0 213-500 501-1000 1001-2000	30 45 55 65 80 95	None
Moisture-, Heat- and Oil-Resistant Thermoplastic	MTW	60°C / 140°F 90°C / 194°F	Machine Tool Wiring in wet locations as permitted in NFPA Standard No. 79 (See Article 670). Machine Tool Wiring in dry locations as permitted in NFPA Standard No. 79 (See Article 670).	Flame-Retardant, Moisture-, Heat- and Oil-Resistant Thermoplastic	22-12 10 8 6 4-2 1-4/0 213-500 501-1000	(A): 30, 45, 60, 60, 80, 95, 110 (B): 15, 20, 30, 30, 40, 50, 60, 70	(A) None (B) Nylon jacket

* Outer covering shall not be required over rubber insulations which have been specifically approved for the purpose.

** For 14-12 sizes RHH shall be 45 mils thickness insulation.

For insulated aluminum and copper-clad aluminum conductors, the minimum size shall be No. 12. See Table 310-18 and 310-19.

Source: National Electric Code, Table 310-13.

FIGURE 5.2. Table of conductor application and insulations. National Electrical Code, Table 310-18. Reprinted, with permission, from NFPA 70, National Electrical Code, Copyright 1977. National Fire Protection Association, Boston, Ma.

Not More Than Three Conductors in Raceway or Cable or Earth (Directly Buried), Based on Ambient Temperature of 30°C (86°F)

Size	Temperature Rating of Conductor. See Table 310–13								Size
AWG MCM	60°C (140°F)	75°C (167°F)	85°C (185°F)	90°C (194°F)	60°C (140°F)	75°C (167°F)	85°C (185°F)	90°C (194°F)	AWG MCM
	TYPES RUW, T, TW, UF	TYPES FEPW, RH, RHW, RUH, THW, THWN, XHHW, USE, ZW	TYPES V, MI	TYPES TA, TBS, SA, AVB, SIS, †FEP, †FEPB, †RHH, †THHN, †XHHW*	TYPES RUW, T, TW, UF	TYPES RH, RHW, RUH, THW, THWN, XHHW, USE	TYPES V, MI	TYPES TA, TBS, SA, AVB, SIS, †RHH, †THHN, †XHHW*	
	AMPS	COPPER				ALUMINUM OR COPPER-CLAD ALUMINUM			
18	. . .	. . .	. . . .	21	. . . .	. . . .	. . . .	. . . .	. . . .
16	. . . .	. . . .	22	22	. . . .	. . . .	. . . .	. . . .	. . . .
14	15	15	25	25	. . . .	. . . .	. . . .	. . . .	. . . .
12	20	20	30	30	15	15	25	25	12
10	30	30	40	40	25	25	30	30	10
8	40	45	50	50	30	40	40	40	8
6	55	65	70	70	40	50	55	55	6
4	70	85	90	90	55	65	70	70	4
3	80	100	105	105	65	75	80	80	3
2	95	115	120	120	75	90	95	95	2
1	110	130	140	140	85	100	110	110	1
0	125	150	155	155	100	120	125	125	0
00	145	175	185	185	115	135	145	145	00
000	165	200	210	210	130	155	165	165	000
0000	195	230	235	235	155	180	185	185	0000
250	215	255	270	270	170	205	215	215	250
300	240	285	300	300	190	230	240	240	300
350	260	310	325	325	210	250	260	260	350
400	280	335	360	360	225	270	290	290	400
500	320	380	405	405	260	310	330	330	500
600	355	420	455	455	285	340	370	370	600
700	385	460	490	490	310	375	395	395	700
750	400	475	500	500	320	385	405	405	750
800	410	490	515	515	330	395	415	415	800
900	435	520	555	555	355	425	455	455	900
1000	455	545	585	585	375	445	480	480	1000
1250	495	590	645	645	405	485	530	530	1250
1500	520	625	700	700	435	520	580	580	1500
1750	545	650	735	735	455	545	615	615	1750
2000	560	665	775	775	470	560	650	650	2000

FIGURE 5.3. Table of allowable ampacities of copper and aluminum conductors. National Electrical Code, Table 310–18.

CORRECTION FACTORS

Ambient Temp.°C	For ambient temperatures over 30° C, multiply the ampacities shown above by the appropriate correction factor to determine the maximum allowable load current.								Ambient Temp. °F
31–40	.82	.88	.90	.91	.82	.88	.90	.91	86–104
41–50	.58	.75	.80	.82	.58	.75	.80	.82	105–122
51–60	....	.58	.67	.71	....	.58	.67	.71	123–141
61–70	....	.35	.52	.58	....	.35	.52	.58	142–158
71–80	....	....	.30	.41	....	....	.30	.41	159–176

†The load current rating and the overcurrent protection for these conductors shall not exceed 15 amperes for 14 AWG, 20 amperes for 12 AWG, and 30 amperes for 10 AWG copper; or 15 amperes for 12 AWG and 25 amperes for 10 AWG aluminum and copper-clad aluminum.

*For dry locations only. See 75°C column for wet locations.

FIGURE 5.3. (Continued) Reprinted, with permission, from NFPA 70, National Electrical Code, Copyright 1977, National Fire Protection Association, Boston, Ma.

conductors unless the circuit is extremely long. For example, if a 5-ton condensing unit pulls 24 amperes, the service mechanic should add 25% of the full-load amperage to the total, which would be 6 amperes. This gives a total of 30 amperes for the wire-sizing data. From the table in figure 5.3, we see that the circuit would require a copper conductor of No. 10 TW copper wire. Also from the table we see that the circuit would require a No. 8 TW aluminum conductor.

Figure 5.4 shows the manufacturer's electrical data for a piece of equipment. The wire sizes are given in columns on the right. Figure 5.5 shows the table of electrical characteristics of a specific model of equipment, but in this table the exact wire sizes are not given. However, the wire size amperages are given (the column headed MWA). The installation mechanic would use the wire size amperage column and the NEC wire charts to size the wire.

There are many items in the NEC that cover the sizing of electric conductors. Most technicians use the NEC for reference. The preceding examples were given as a guide to show you how to use the NEC charts.

Calculating Voltage Drop

The table shown in figure 5.3 does not consider voltage drop, which must be calculated for long circuits. To calculate the voltage drop in a conductor, you must know the number of feet of wire that is used. The following formula is used to calculate voltage drop:

$$E = IR$$

For example, if No. 12 TW wire is to run 500 feet, what is the voltage

BRANCH CIRCUIT

Power Wire Size (AWG)	Max Ft Wire	Gnd Wire Size† (AWG)	Max Fuse Amps
10	40	10	35
10	35	10	40
8	43	10	50
8	35	10	60
8	44	10	60
6	56	8	70
12	32	12	30
10	47	10	35
10	36	10	45
8	45	10	50
8	56	10	50
8	47	10	60
10/12	50/42	10/12	35/30
10/10	45/60	10/10	35/30
10	53	10	40
10	45	10	45
10	68	10	35
10	57	10	40
14	79	12	20
14	67	12	20

FIGURE 5.4. Table of recommended wire sizes for an air conditioner. Reproduced by permission of Carrier Corporation. © 1977 Carrier Corporation.

†Required when using nonmetallic conduit.

drop over this distance when the supply voltage is 240 volts? First, we find the resistance of No. 12 wire from figure 5.1. It is 1.6 ohms per 1000 feet. From figure 5.3 we find the ampacity of No. 12 TW wire. It is 20 amperes. Substituting in the formula $E = IR$, we find $E = 20 \times .8$ (1.6 ohms per 1000 feet = .8 ohm per 500 feet). The voltage drop over 500 feet is 16 volts. Subtracted from the supply voltage of 240 volts, this gives 224 volts supplied to the equipment.

The formula

$$E_d = \frac{22 \times D \times I}{cmil}$$

can also be used, where E_d is the voltage drop, D is the distance, I is the current, and cmil is circular mils.

In most cases the wiring that is run to heating, cooling, or refrigeration equipment will not exceed 75 feet to 100 feet. Then the wire size can be read directly from the table in figure 5.3 because the voltage drop can be ignored on short-distance circuits. (There is a correction factor in the NEC tables if the temperature exceeds 30°C.)

| UNIT MODEL 38AD | V/PH | | | | | COMPR | | FANS FLA (ea)* |
	Nom	Min-Max	MWA	ICF	FU	FLA	LRA	
410	208/3	187-229	68	198	80	49.3	191	3.2
012 510	230/3	198-254	62	179	70	44.3	172	3.2
610	460/3	414-506	31	89	35	22.2	86	1.6
110	575/3	518-632	24	71	30	17.9	69	1.0
400	208/3	187-229	86	184	100	63.6	173/266	3.2
014 500	230/3	198-254	78	166	90	57.2	160/240	3.2
600	460/3	414-506	39	123	45	28.6	120	1.6
100	575/3	518-632	32	99	35	22.9	96	1.0
400	208/3	187-229	86	184	100	64.0	173/266	3.2
016 500	230/3	198-254	78	166	90	58.0	160/249	3.2
600	460/3	414-506	39	123	45	29.0	96	1.6
100	575/3	518-632	32	99	40	23.0	96	1.0

☐   Values are for part-winding start.

FLA — Full Load Amps
FU — Fuse (max allowable amps; dual element)
ICF — Max Instantaneous Current Flow (during start-up; sum of compressor LRA plus FLA of all other motors in the unit).
LRA — Locked Rotor Amps
MWA — Minimum Wire Amps per NEC

*Units 38AD012 and 014 have two; unit 38AD016 has three.

FIGURE 5.5.   Table of electrical characteristics of a large condensing unit. Reproduced by permission of Carrier Corporation. © 1977 Carrier Corporation.

5.2   DISCONNECT SWITCHES – *close to load for safety fused or not fused*

All heating, cooling, and refrigeration equipment should have some means for disconnecting the power supply at the equipment. Some equipment has a built-in method for disconnecting the power, such as a circuit breaker or fuse blocks. However, in most cases a disconnecting device must be supplied and installed by an electrician or the installation mechanic. Disconnect switches are relatively simple and easy to install once the correct selection is made.

A disconnect switch is a two- or three-pole switch mounted in an enclosure. The switch can be purchased with or without a space for fuses. Disconnect switches in most cases have a grounding lug mounted in the enclosure. The switches can have different arrangements. For example, a four-pole–three-fuse switch would be used on a three-phase circuit and would have a grounding lug. A three-pole–three-fuse switch would be used for three-phase circuits and would not have a ground lug. A three-pole–two-fuse switch would be used on single-phase circuits and would have a ground lug. Figure 5.6 shows the schematics of two disconnect switches and figure 5.7 shows these switches.

Disconnect switches can be purchased for general duty or heavy duty.

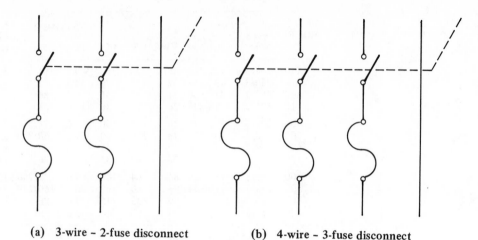

(a) 3-wire – 2-fuse disconnect   (b) 4-wire – 3-fuse disconnect

FIGURE 5.6. Schematic diagram of two disconnect switches.

The heavy-duty disconnect switch would be installed for equipment that would require frequent use. The general-duty switch would be used for equipment requiring infrequent use.

Enclosures *for disconnect switch*

The type of enclosure that the disconnect switch mounts in is determined by the conditions existing in the area of installation. A general type of enclosure could only be used where there were no problems of moisture, dust, or explosive fumes. A watertight disconnect enclosure could be used in areas of moisture but not where dust or explosive fumes exist. An explosionproof enclosure could be used in any location, but it is much more expensive than other enclosures and hence is not used without reason.

Fusible and Nonfusible Switches

The purpose of a disconnect switch can be twofold. First, it can be used as a means of disconnecting the supply power going to the equipment. Second, it can be used as a safety device when fused correctly. If the only purpose of a disconnect switch is to break the power supply, then a nonfusible disconnect switch should be used. But if a means of protection for the wire or equipment is needed, a fusible disconnect switch should be used with the proper size fuses. Most equipment manufacturers will give the fuse sizes needed in

(a) 3-wire – 2-fuse disconnect

(b) 4-wire – 3-fuse disconnect

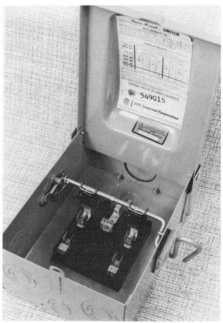

FIGURE 5.7   Disconnect switches. Photo courtesy of Gould I-T-E Electrical Products.

the installation instructions. If fuse sizes are not given, the *National Electrical Code* should be consulted.

The selection of a fusible disconnect switch is determined by duty, enclosure type, and size. Fuses are designed so that one size covers several different ampacities. The same size fuse can be purchased to cover from 1 to 30 amperes, from 30 to 60 amperes, from 70 to 100 amperes, and from 100 to 200 amperes. There are larger sizes available, but they are not used frequently.

Disconnect switches are rated 30 amperes, 60 amperes, 100 amperes, 200 amperes, 400 amperes, and 600 amperes. A 30-ampere disconnect switch would be used for any load from 1 ampere to 30 amperes. A 200-ampere disconnect switch can be used with fuses from 100 amperes to 200 amperes. Other determining factors of the switches can easily be selected from a manufacturer's catalog.

Figure 5.8 shows a disconnect switch installed on a piece of equipment.

5.3 FUSIBLE LOAD CENTERS ~ ~~BREAKER~~ ~~PANEL~~ 60 A

Fusible load centers, or breaker panels, are electric panels that supply the circuits in a structure with power and protect those circuits with fuses. Figure 5.9 shows a typical fusible load center. Fusible load centers were popular until about 1968, when breaker panels gained prominence in the market. Air-conditioning mechanics often find themselves working on fusible load centers when the owner of an older home decides to have air conditioning installed. There are many fusible load centers still in existence, so mechanics must understand them and know how to make the correct connections. In many cases it is impossible to add a 230-volt circuit to a fusible load cen-

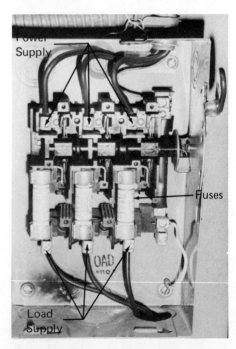

FIGURE 5.8. Disconnect switch installed on equipment. Photo courtesy of General Electric Co., Circuit Protective Devices Products Dept.

FIGURE 5.9. Fusible load center. Photo courtesy of Crouse-Hinos Company, Distribution Equipment Div.

ter, so the service technician must know when to recommend installing a new breaker panel.

One of the earliest fusible load centers is shown in schematic form in figure 5.10. The panel had a 60-ampere lighting main, which means that the total lighting or 110-volt circuits in the residence could not exceed 60 amperes on the eight circuits. There was also a 60-ampere–230-volt circuit for a kitchen stove. The addition of any electric circuit to the 60-ampere fusible load center is almost impossible because of its construction and its low capacity. Most modern residences require a 150-ampere or 200-ampere service. Fusible load centers built after about 1958 are capable of handling the load.

There have been many fusible load centers installed in residences that are much better and more versatile than the original 60-ampere centers. Many are designed with a lighting main and a capacity to take a number of additional 230-volt circuits.

The newest fusible load centers are built with the capacity of taking additional fuse blocks in the main lugs, which supply power to the entire panel, at any time if a space is available. This capacity allows the electrician to install the additional circuits required for any particular structure. Figure 5.11 shows this panel and its schematic. It is relatively easy to add a circuit

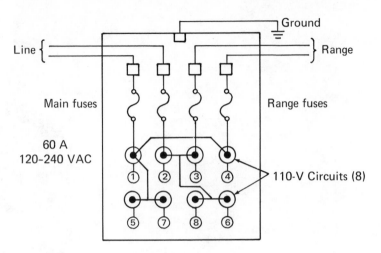

FIGURE 5.10.   Schematic of an early fusible load center.

to the late-model panels by merely inserting the fuse blocks and screwing them in.

5.4 BREAKER PANELS

Breaker panels are usually installed in residences and industrial buildings to-day. Breakers are devices that detect any overload above their rating in a circuit and open the circuit automatically. The breaker must then be reset manually.

Breakers can be obtained for almost any application, no matter how large or small. Breakers are made with one, two, or three poles. The poles denote how many hot legs are being fed from the breaker to the appliance. A one-pole breaker supplies one hot leg and makes up a 110-volt–single-phase circuit. A two-pole breaker supplies two hot legs and makes a 230-volt–single-phase circuit. A three-pole breaker supplies three hot legs and makes up a three-phase circuit at the supplied voltage. Breakers are supplied in amperage ratings of one ampere to several hundred amperes, depending on the application. Figure 5.12 shows one-pole, two-pole, and three-pole breakers.

Construction

Breaker panels are built in several different forms, but they all serve the same purpose. Different manufacturers build breaker panels that are similar in de-

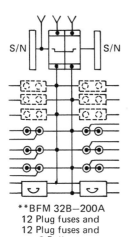

**BFM 32B—200A
12 Plug fuses and
12 Plug fuses and
2 Pullouts

(a)   Schematic

Drawing Guide

Available = circuit = space	4 = Plug = fuses	1 Fuse pullout
Available = circuit = space		1 Fuse pullout

FIGURE   5.11.   Modern   fusible   load center.

(b)   Load center

FIGURE 5.12. Single-pole (a), double-pole (b), and triple-pole (c) breakers. Photos courtesy of General Electric Co., Circuit Protective Devices Products Dept.

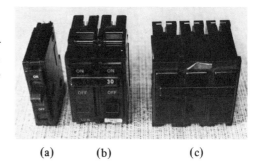

(a)   (b)   (c)

113

sign, but the breakers from different manufacturers do not usually fit each others' panels. However, several manufacturers' equipment does interchange.

Breaker panels are built with or without main breakers. A main breaker installed in the breaker panel provides a main switch in the panel and adds a means of overload protection for the entire panel. A breaker panel can be obtained with main lugs and no main breaker, but the breaker should have some means of protection. Breaker panels are rated by how many amperes the main lugs can carry and by the rating of the main breaker.

Breaker panels are built for use with single-phase or three-phase systems and for 250 volts or 600 volts. The breaker panel in the average residence is rated at 150 amperes or 200 amperes and is a general-duty type. The breakers snap into a residential panel (as they do in some commercial and industrial panels). Figure 5.13 shows a residential breaker panel and its schematic.

The commercial and industrial breaker panel can meet almost any specifications that the consumer requires. The commercial and industrial panels are built for more rugged duty than are the usual residential panels. Most commercial and industrial panels use breakers that attach to main lugs with screws. Figure 5.14 shows a typical breaker panel, along with its schematic, used in commercial and industrial applications.

Installation

Installation of a breaker into a breaker panel causes little or no trouble. Breakers connect in some panels by attaching to the main lugs with clips.

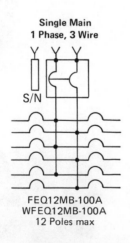

Single Main
1 Phase, 3 Wire

S/N

FEQ12MB-100A
WFEQ12MB-100A
12 Poles max

(a) Schematic

(b) Panel

FIGURE 5.13. Residential breaker panel. Photo courtesy of Gould I-T-E Electrical Products.

FIGURE 5.14. Industrial breaker panel. Schematic courtesy of Gould I-T-E Electrical Products.

>─▭─ S/N
EQ 430MBD—400A
EQ 430MBE—600A
30 poles max.

(a) Schematic   (b) Panel

Other breakers are attached to the main lugs by screws. You should be familiar enough with breakers and breaker panels to order the correct breaker for a particular use.

In almost all cases breaker panels have several open spaces. If a situation arises where there is no spare opening, a smaller breaker can replace the larger standard breakers. Most manufacturers provide the smaller breaker. Figure 5.15 shows the comparison between the standard breaker and the smaller compact breaker.

On rare occasions circuit breakers are faulty. Either they cannot be reset or they open the circuit on a lower amperage than the rating. If either condi-

FIGURE 5.15. Standard breaker and small replacement breaker. Photo courtesy of General Electric Co., Circuit Protective Devices Products Dept.

tion exists, the breaker must be replaced. A breaker can be checked for re-setting if a voltage reading is taken between the ground and the breaker. If voltage is available to the load side of the breaker under load, the breaker is good. In rare cases a breaker could be stuck in the closed position, and if so, it should be replaced. The breaker that trips at a lower-than-rated amperage should be checked with an actual ampere reading of the circuit.

5.5 DISTRIBUTION CENTERS

Distribution centers are designed to distribute the electrical supply to several places in a large structure. Their use is largely confined to commercial and industrial applications. Figure 5.16 shows schematically the layout of a distribution system in a structure. In figure 5.16 panel A is the main distribution point between the electrical service and other panels and the heating and air-conditioning equipment.

Figure 5.17 shows a modern distribution panel. In a very large structure

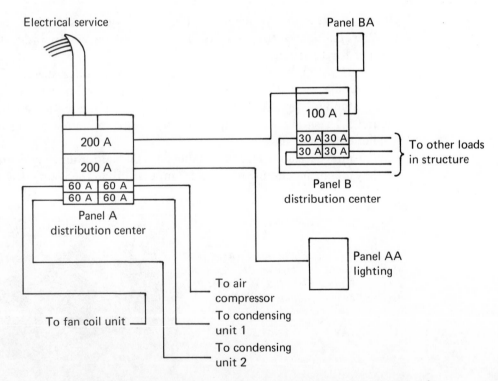

FIGURE 5.16. Schematic of the electrical distribution system of a large structure.

FIGURE 5.17. Modern distribution panel. Photo courtesy of General Electric Co., Circuit Protective Devices Products Dept.

it would not be unusual to find several distribution centers. A distribution center often saves a great deal of money because many circuits are shortened and only the main circuit is lengthy. Distribution centers can be of the fusible or circuit-breaker design. The fusible design is the more popular mainly because of its moderate cost. Larger breakers are extremely expensive and hence are not used often.

SUMMARY

For proper installation and maintenance of heating, cooling, and refrigeration equipment, industry personnel should be familiar with the electric circuits and circuit components of a structure. The life and safety of the equipment depend on the use of the correct size of wiring. Voltage drop, insulation type, enclosure, and safety are the determining factors in wire sizing. Manufacturers usually provide instructions on the size of wire to use with their equipment, but in some cases the mechanic must calculate the size. The *National Electrical Code* is a guide that should be used in sizing wire properly.

There are various types of electric panels used in the industry. The disconnect switch, sometimes called the safety switch, is commonly used on equipment, along with another electric panel. In almost all structures the electric panels are not located close enough to the equipment to be considered safe for disconnecting the equipment. Therefore, in most cases disconnect switches should be installed on or close to the equipment.

The fusible load center or breaker panel is used in almost all residences. The breaker panel is also popular in commercial and industrial panels. Breakers are designed to trip or break the circuit on an overload. Some breakers clip to the main lugs; others screw to them.

The distribution center is used in commercial and industrial plants as a means of distributing power to other electric panels in the structure. Service personnel should be familiar with power distribution because of the many voltage ranges of modern equipment. The mechanic or installer must be able to pick up power out of any type of electric panel. In many areas industry personnel are responsible for the total installation of equipment, including power wiring. In other areas power wiring must be done by an electrician.

QUESTIONS

1. What size wire could be used for a circuit requiring 100 amperes? *1*

2. What are the four factors that should be considered in sizing circuit conductors?

3. What is a disconnect switch?

4. What are the three basic types of disconnect switch enclosures?

5. What is a fusible load center?

6. What is a circuit breaker?

7. How can a faulty circuit breaker be detected?

8. How are circuit breakers attached to the lugs of an electric panel?

9. What is a distribution center and what is its primary use?

10. The most popular conductor used in the industry is _____.

11. True or false. The American Wire Gauge lists the standard wire sizes used in the United States today.

12. What is the major disadvantage of aluminum wiring?

13. How would you determine the voltage drop in a circuit?

14. For wire-sizing problems the service technician should consult the _____.

15. What is the purpose of a disconnect switch?

16. True or false. Fusible load centers are usually installed in new homes today.

17. Breaker panels are made with _____ poles. The poles denote _____.

18. What is the purpose of a main lug in a breaker panel?

19. What would you do in a situation where there is no spare opening in a breaker panel for installation of equipment?

20. If a 10-TW wire is 750 feet, what is the voltage drop in 230 volts?   *.765· OHMS*

21. What would the supply voltage be for a 100-foot 4/0 wire if 150 amperes are used on 230 volts?

ON FINAL

6

Basic Electric Motors

INTRODUCTION

The electric motor changes electric energy into mechanical energy. Motors are used to drive compressors, fans, pumps, dampers, and any other device that needs energy to power its movement.

There are many different types of electric motors with different running and starting characteristics. Most single-phase motors are designed and used according to their running and starting torque. **Torque** is the strength that a motor produces by turning, either while starting or running. This chapter covers most types of motors available today and how they are used in the heating, cooling, and refrigeration industry.

We begin our study with a discussion of magnetism, an effect that is needed to operate motors, relays, contactors, and other electric devices.

6.1 MAGNETISM

Magnetism is the physical phenomenon that includes the attraction of an object for iron and is exhibited by a permanent magnet or an electric current. Magnetism is produced in many different ways, but regardless of how it is produced, the effect is basically the same. The magnetic field of the earth, for example, is the same as the magnetism in a horseshoe magnet, the magnetism produced by a transformer, and the magnetism produced by an electromagnet. A good example of magnetism is the ability of a horseshoe magnet

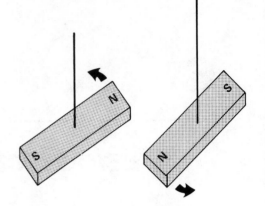

FIGURE 6.1. Repulsion of like poles
of a magnet.

to pick up articles made of iron. The most common example of magnetism is the reaction of a compass to the earth's magnetic field.

All magnets have two poles, a north pole and a south pole. If the north pole of a bar magnet is brought close to the north pole of another bar magnet, they will repel each other, as shown in figure 6.1. If the south pole of one bar magnet is brought close to the north pole of another bar magnet, they will attract each other and come together, as shown in figure 6.2. Therefore, like poles of magnets repel each other and unlike poles attract.

Magnetic Field

The magnetic lines of force of a magnet that flow between the north and south poles are called **flux**. These lines of force are shown in figure 6.3. The area that the magnetic force operates in is called a **magnetic field**. Magnetic fields can flow through materials, depending on the strength of the magnetic field. A magnetic field is best conducted through soft iron. That is why certain parts of motors and other electric devices are made of soft iron.

FIGURE 6.2. Attraction of unlike poles
of a bar magnet.

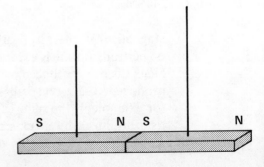

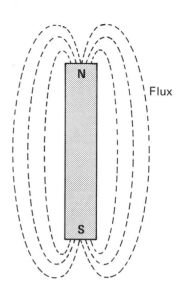

FIGURE 6.3. The magnetic field of a
bar magnet.

Induced Magnetism

<u>Induced magnetism</u> is created when a piece of iron is placed in a magnetic
field. The important fact to remember about a magnetic field is that the
closer an object is to the magnet, the stronger the magnetic field is on that
object. Therefore, if we insert an iron bar within two or three inches of a
magnetic field, we induce a stronger field than if we placed the bar six inches
from the field.

There are two types of magnets in use today: the permanent magnet and
the electromagnet. The **permanent magnet** is a piece of magnetic material
that has been magnetized and can hold its magnetic strength for a reasonable
length of time. The permanent magnet must be made of a magnetic material,
such as iron, nickel, cobalt, or chromium. Some nonmagnetic materials such
as glass, rock, wood, paper, and air cannot be made magnetic but can be
penetrated by a magnetic field.

The **electromagnet** is a magnet that is produced by electricity. When
there is an electron flow in a conductor, a magnetic field is created around
the conductor, as shown in figure 6.4. The larger the electron flow, the
stronger the magnetic field. Therefore, if we take an iron core and wind a
current-carrying conductor around it, the iron core will become a magnet, as
shown in figure 6.5. The electron flow and the number of turns of the conductor around the core determine the strength of an electromagnet. Figure
6.6 shows an electromagnet that is used as a solenoid in a contactor.

Magnetism is important in the heating, cooling, and refrigeration industry
because of its many uses in the operation of electric devices. Motors require

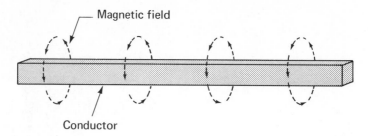

FIGURE 6.4. The magnetic field created around a current-carrying conductor.

magnetism to create a rotating motion. Relays and contactors use magnetism to open and close a set of contacts. All the devices discussed in this chapter use magnetism in some way.

6.2   THE BASIC ELECTRIC MOTOR

Electric motors are very common electric devices in the heating, cooling, and refrigeration industry. Motors are used to create a rotating motion and drive components that need to be turned. Motors power compressors, pumps, fans, timers, and any other device that must be driven with a rotating motion.

In an electric motor, electric energy is changed to mechanical energy by magnetism, which causes the motor to turn. The method by which magnetism causes motors to rotate uses the principle that like poles of a magnet repel and unlike poles attract. Suppose a simple magnet is placed on a pivot and used as a **rotor** (the rotating part of an electric motor) and a horseshoe magnet is used as a **stator** (the stationary part of a motor), as shown in figure 6.7(a). Movement will be obtained by the repulsion and attraction of the

FIGURE 6.5. The magnetic field of an iron core when a current-carrying conductor is wound around the core.

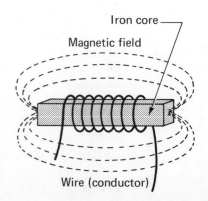

122

FIGURE 6.6. An electromagnet used as a solenoid in a contactor. Photo courtesy of Essex Group, Controls Div.

poles of a magnet. The rotor would turn until the unlike poles are attracted to each other, as shown in figure 6.7(b).

To make an electric motor move continuously, we must have a rotating magnetic field, which is produced by the reversal of the poles, or the polarity, in the rotor or stator. An alternating current of 60 hertz changes direction 120 times per second. Therefore, the current would change the polarity of the stator poles on each reversal of current. If the rotor has a permanent polarity, as shown in figure 6.8, then the changes of polarity in the stator would cause the rotor to move. Therefore, if alternating current changes direction, causing a polarity change, 120 times a second, then the motor will turn in a continuous motion because the poles of the stator will be continuously repelling and attracting the permanent poles of the rotor. Figure 6.8 shows the

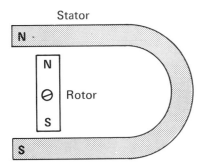

(a) **Initial position of the rotor**

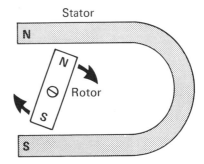

(b) **Movement of the rotor due to the repulsion and attraction of the magnets' poles**

FIGURE 6.7. A simple electric motor.

Two-pole motor, 3600 rpm

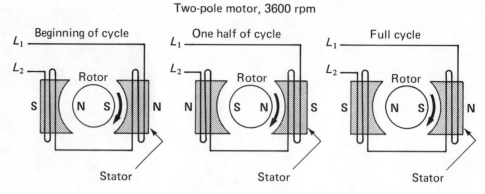

FIGURE 6.8.   A complete cycle of operation of an electric motor.

motor in one complete cycle of current, or $\frac{1}{60}$ second. The movement of the motor is caused by the magnetic field of the stator as it rotates through its alternations of current. Figure 6.9 shows the coils of an ordinary electric motor and its windings. The coils of the motor are a part of the stator.

In motors, the rotor is not a permanent magnet, as we stated in the previous explanation. The **squirrel cage rotor** shown in figure 6.10 is the most commonly used rotor today. The squirrel cage rotor derives its name from its cagelike appearance. In the squirrel cage rotor, copper or aluminum bars are evenly spaced in the steel portion of the rotor and connected by an aluminum or copper end ring. The squirrel cage rotor produces an inductive magnetic field within itself when the stator is energized.

The most common motors operate much like a transformer, with the stator being the primary magnetic field and the rotor being a movable secondary magnetic field. The rotor will have magnetism induced into it from the stator and its magnetic poles will be permanent. The magnetic poles of the stator are moving at the rate of the alternations of the current.

FIGURE 6.9.   The windings of an electric motor.

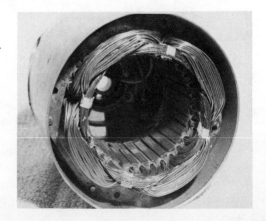

FIGURE   6.10.   Squirrel   cage   rotor.

6.3 TYPES OF ELECTRIC MOTORS

The industry uses all kinds of AC motors to rotate the many different devices that require rotation in a complete system. Different motors are needed for different tasks because all motors do not have the same running and starting characteristics. This fact, along with the increased cost of stronger motors, allows the industry to use the right motor for the right job. For example, compressors require a motor with a high starting torque and a good running efficiency. Small propeller fans use motors with a low starting torque and average running efficiency.

Motor Strength

The starting methods, or strength, are generally used to classify motors into types. Motors are selected mainly because of the starting torque (power) that is required for the motor to perform its function. There are basically five general types of motors used in environmental systems: shaded-pole, split-phase, permanent split-capacitor, capacitor-start–capacitor-run, capacitor-start, and three-phase. There are others, such as the repulsion-start–induction-run and series motors. However, these are outdated or not commonly used in the industry. The starting torques of the five general types of induction motors, expressed as a percentage of their running torque, are as follows: shaded-pole, 100%; split-phase, 200%; permanent split-capacitor, 200%; capacitor-start–capacitor-run, 300%; and three-phase, 600%.

Open and Enclosed Motors

Two other motor types are the open motor and the enclosed motor. The open motor, shown in figure 6.11, has a housing and is used to rotate a de-

125

FIGURE 6.11. Open motor. Robbins & Myers, Electric Motor Div.

vice such as a fan or a pump that is itself not enclosed in any type of housing. The enclosed motor is housed within some type of shell. The most common enclosure of a motor is a completely sealed hermetic compressor, as shown in figure 6.12. Any starting apparatus used on an enclosed motor must be mounted outside the enclosure. The starting apparatus of an open motor is usually mounted within the motor itself.

In the following sections we will discuss the five basic motor types in detail.

6.4 SHADED-POLE MOTORS

A **shaded-pole motor,** shown in figure 6.13, is an induction-type motor that does not incorporate an ordinary starting winding. It uses a band on one

FIGURE 6.12. Enclosed motor used in hermetic compressors.

FIGURE 6.13. Shaded-pole motor.

side of each pole to get a short-circuit effect that produces a rotating magnetic field to start the rotation. Shaded-pole motors are used when very small starting and running torques are required, such as in a furnace fan, a small condensing unit fan, and an open-type fan. These motors are easily stalled, but in most cases, because of the small locked rotor amperes (i.e., the current draw of the motor when power is applied but the motor does not turn), they can stall and still not burn out the windings.

Operation

Figure 6.14 shows the stator section of a shaded-pole motor. At one side of each pole a small groove has been cut into the stator. These grooves are the shaded pole and are banded by some type of metal, usually copper. This band actually separates the magnetic field created in the stator. The separation makes the shaded-pole polarity differ from the polarity of the main winding, causing the motor to rotate.

In the stator of a shaded-pole motor, a shifting of the magnetism in the poles and in the shaded poles causes a rotating magnetic field in the stator.

FIGURE 6.14. Stator of a shaded-pole motor.

These interactions produce rotation. The unbanded section of the motor stator has the stronger magnetic field at the beginning of the cycle. The magnetic field then builds up on both the main poles and the shaded poles together, and then in the shaded pole, which causes the rotation.

Reversing

Shaded-pole motors are difficult to reverse because to do this they must be disassembled. The rotation of the shaded-pole motor is determined by the location of the shaded pole. Figure 6.15 shows a layout of a shaded-pole motor. If the shaded pole is on the left side of the main pole, as it is in figure 6.15, the rotation will be toward the shaded pole, or clockwise. On the other hand, if the shaded pole is on the right side of the main pole, the rotation will again be toward the shaded pole, but in this case the rotation will be counterclockwise. Therefore, to reverse the shaded-pole motor, the stator must be reversed to change the positions of the shaded poles, and this usually means disassembling the motor.

Troubleshooting

Shaded-pole motors are easy to identify because of the copper band around the shaded pole. The motors are easily diagnosed for trouble because of their simple winding patterns, as shown schematically in figure 6.16. The shaded-

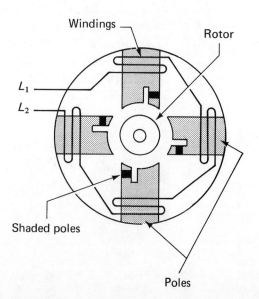

FIGURE 6.15. Layout of a shaded-pole motor.

FIGURE 6.16. Schematic wiring diagram of the shaded-pole motor.

pole motor can be checked with an ohmmeter to determine the condition of
the windings. Just because a shaded-pole motor has been stalled does not
mean that the windings are faulty. If this condition should occur, the motor
probably needs lubrication. The shaded-pole motor is simple and easy to
troubleshoot. It is used in many applications in the industry.

6.5 SPLIT-PHASE MOTORS

The **split-phase motor** shown in figure 6.17 has both a starting winding to
assist the motor in starting and a running winding to continue rotation after
the motor has reached a certain speed. Most single-phase motors have some
method of beginning the rotation; in a split-phase motor, rotation is started
by splitting the phase to make a two-phase current. The single-phase current
is split between the running and the starting winding, which puts one of the
windings out of phase by 45 to 90 degrees. The starting windings are used to
assist the split-phase motor in starting. They are also used until the motor
has reached a speed that is about 75% of its full-capacity speed. The starting
windings then drop out of the circuit by use of a centrifugal switch. After
that occurs the motor operates at full speed on the main or running winding
alone.

A cutaway view of a capacitor-start–capacitor-run motor is shown in
figure 6.18. Without the capacitors it closely resembles a split-phase motor.

The split-phase motor can be operated on 110 volts–single phase–60 hertz
or 208/230 volts–single phase–60 hertz. Some split-phase motors can operate

FIGURE 6.17. Split-phase motor. Photo
courtesy of Emerson Electric Co., Emerson Motor Div.

FIGURE 6.18. Cutaway view of an electric motor. Photo courtesy of Gould, Inc., Electric Motor Div.

on either voltage by making simple changes in their wiring if so desired. Thus it is a dual-voltage motor. Split-phase motors can be reversed by reversing the leads of the starting winding at the terminals in the motor.

Split-phase motors are used when a high starting torque is not required. They are used in such equipment as belt-driven evaporator fan motors, hot-water pumps, small hermetic compressors grinders, washing machines, dryers, and exhaust fans.

Operation

The phases in a split-phase motor are split by the makeup of the starting windings. The starting winding is designed with smaller wire and more turns than is the running winding, which has a greater inductance. Therefore the running winding is displaced from the starting winding because of its greater inductance. This displacement causes a resistance to current flow to build up in the running winding. The phase displacement means the current reaches the two windings at separate times, allowing one winding to lead, in this case the starting winding. However, some manufacturers allow current to reach the running winding first by designing an increased resistance into the starting winding and a decreased induction into the running winding. But whatever method is used, the motor basically operates on the same principle: splitting the phase.

The operation of a split-phase motor, referring to figure 6.19, is as follows:

1. Power is applied to the running and starting windings in parallel. The motor itself splits the phase by the counter electromotive force (emf) in the starting winding, which acts as a resistance to

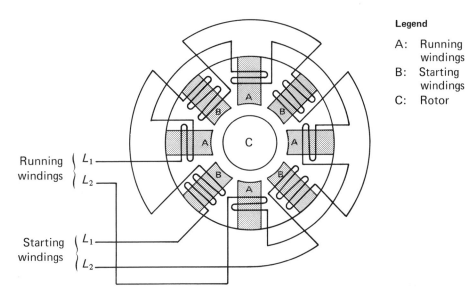

Legend

A: Running windings
B: Starting windings
C: Rotor

Running windings { L_1, L_2

Starting windings { L_1, L_2

FIGURE 6.19.   Layout of a split-phase motor.

hold back the current flow to the starting winding. On the first alternation of power, the running winding creates a higher magnetic field than the starting winding.

2. In half of a cycle the alternations are changed. The starting winding has the stronger magnetic field, moving the rotor a certain distance depending on the number of poles in the motor. For the motor in figure 6.19 the distance is one-fourth of a rotation.

3. As the alternations continue at the rate of 60 cycles per second, the motor continues to rotate by the stator. Therefore, the rotor with its magnetic field attempts to keep up with the rotating magnetic field of the stator.

4. The motor is equipped with a centrifugal switch that draws the starting winding out of the circuit when the motor has reached 75% of its full speed.

Troubleshooting

Split-phase motors are one of the most reliable motors used in the heating, cooling, and refrigeration industry. They are used on almost all types of single-phase equipment. The split-phase motor is fairly easy to troubleshoot if the service mechanic has a good understanding of its operation. The three probable areas of trouble are the bearings, the windings, and the centrifugal switch.

FIGURE 6.20. Centrifugal switch used
in a split-phase motor.

The bearings of any type of motor often give trouble because of wear and improper maintenance. Identification of a motor with bad bearings is simple. The motor will have trouble turning and in some cases may be locked down completely.

The windings of a split-phase motor can be shorted, open, or grounded. This is easily diagnosed with an ohmmeter.

The centrifugal switch, shown in figure 6.20, is the hardest section to diagnose for troubles because it stays in the circuit only a short time. The centrifugal switch has a tendency to stick in an open or closed position because of wear and it often must be replaced. The centrifugal switch can usually be heard when it drops in after the motor is cut off. Hence it can be checked effectively in this manner. If the centrifugal switch does not drop out of the circuit, the motor will pull an excessively high ampere draw and cut off on overload. One of the surest methods of checking a centrifugal switch is by disassembling the motor and making a visual inspection.

Before presenting the other motor types, we turn to a discussion of a special electric device, the capacitor, which plays an important role in those motors.

FIGURE 6.21. Symbol for a capacitor.

6.6 CAPACITORS

The **capacitor** consists of two aluminum plates with an insulator between them. The insulator prevents electrons from flowing from one plate to the other, but it allows storage of electrons. Figure 6.21 shows the schematic symbol for a capacitor. Capacitors are used to boost the starting torque or running efficiency of single-phase motors.

Two Types Used in the Industry

Two types of capacitors are used primarily in the industry: the electrolytic, or starting, capacitor and the oil-filled, or running, capacitor (see figure 6.22). Starting capacitors consist of two aluminum electrodes (plates) with a chemically treated paper, impregnated with a nonconductive electrolyte, between them. They can be purchased in ranges from 75 to 600 microfarads (μF) and

(a) Starting capacitor

(b) Running capacitor

FIGURE 6.22. Two capacitors used in the industry.

from 110 to 330 volts. A **microfarad** is the unit of measurement for the strength of a capacitor; all capacitors are rated according to their strength in microfarads. The electrolytic capacitor is used to assist a single-phase motor in starting. *(RUN)*

The oil-filled capacitor consists of two aluminum electrodes with paper between them and an oil-filled capacitor case. It is available in microfarad ranges of about 2 to 60 and voltage ranges of 230 to 550. The oil-filled capacitor can be used for small or moderate torque starting, but it is more commonly used to increase a motor's running efficiency.

The major difference between the two types of capacitors is in their purpose. A starting capacitor is built in a relatively small case with a **dielectric**—a nonconductor of electric current. It is used for only a short period of time on each cycle of the motor. Therefore, a starting capacitor has no need to dissipate heat, although its capacity is larger than its counterpart, the running capacitor.

The **running capacitor** is designed to stay in the motor circuit for the entire cycle of operation. Therefore it must have some means of dissipating the heat. The oil in the capacitor case is used for this purpose. The oil-filled capacitor is physically larger than the starting capacitor but smaller in capacity than the starting capacitor. Both capacitors are in wide use in the industry and serve two very basic needs.

Troubleshooting

Short capacitor life and malfunctions can be caused by several different factors. High voltage can cause a capacitor to overheat. This can damage the plates and short the electrodes. Starting capacitors can be damaged by faulty starting apparatus that would keep the capacitor in the line circuit long enough to damage the capacitor. Excessive temperature can shorten the life of capacitors or cause permanent damage due to poor ventilation, starting cycles that are too long, or starting cycles that occur too frequently. The cause of the malfunction should be corrected as soon as possible. The capacitors themselves are frequently the cause of the problem.

All capacitors used on single-phase motors are designed specifically to assist the motor in proper operation. However, in some cases it is impossible to replace a capacitor with an exact replacement. If this situation should occur, use the following guidelines for replacing the capacitor:

1. The voltage of any capacitor used for replacement must be equal to or greater than the capacitor being replaced.
2. The strength of the starting capacitor replacement must be at *(μF)*

least equal to but not greater than 20% of the capacitor being replaced.

3. The strength of the running capacitor replacement may vary a plus or minus 10% of the capacitor being replaced.

4. If capacitors are installed in parallel, the sum of the capacitors is the total capacitance.

5. The total capacitance of capacitors in series may be found by the following formula:

$$C = \frac{C_1 C_2}{C_1 + C_2}$$

These rules are intended only as a guide. Remember that it is always preferable to use an exact replacement.

There are many methods for testing capacitors in common use in the industry today. A capacitor can be checked by using an ohmmeter. The ohmmeter should be placed on a high-ohm scale and both leads should be connected to the terminals of a discharged capacitor. If the needle of the meter shows a deflection to the right end of the scale and back to zero, the capacitor is good. If the needle comes to rest on 0 ohms, the capacitor is shorted. If the needle of the meter does not move, it indicates an open capacitor.

In case of doubt there is another method that can be used to check the capacitor. By briefly applying voltage to a capacitor, reading the amperage, and then substituting the values into the formula

$$\text{microfarads} = \frac{2650 \times \text{amperes}}{\text{volts}}$$

the exact capacitance can be obtained. There are also many commercial capacitance testers available on the market.

6.7 PERMANENT SPLIT-CAPACITOR MOTORS

Permanent split-capacitor motors (PSC) are simple in design and have a moderate starting torque and a good running efficiency, which makes them a popular motor in the industry. Figure 6.23 shows a PSC motor that is used as a fan motor. The starting winding and the running capacitor of the PSC motor are connected in series, as shown in figure 6.24(a). The schematic diagram of the hookup is shown in figure 6.24(b). The running and starting windings are in parallel, but the capacitor causes a phase displacement.

Permanent split-capacitor motors are used on compressors, where the

FIGURE 6.23. Permanent split-capacitor motor used as a fan motor. Photo courtesy of Universal Electric Co.

refrigerant equalizes on the "off" cycle, on direct-drive fan motors, and on most residential air conditioners of 5 horsepower or under. It has a relatively low cost in comparison with other motors because it does not have a switch to drop the starting winding. The PSC motor can be used only when a moderate starting torque is required to begin rotation.

Operation

Permanent split-capacitor motors operate much like the split-phase motor, but the starting winding stays energized for the total cycle. The phase displacement of the motor is created by the running capacitor, which stores electrons momentarily. The capacitor causes the starting winding to lead the running winding and thus allows the motor to start rotation.

Troubleshooting

The PSC motor usually gives trouble-free operation for long periods of time. The three most common failures that occur in a PSC motor are in the bearings, windings, or a capacitor.

The bearings of a PSC motor often become faulty because of wear or lack of proper maintenance. Bearings in any type of motor can be diagnosed with little trouble by rotating the motor by hand and noticing rough places in the movement or the shaft being locked in one position.

The windings of a motor become faulty because of overheating, overloading, or a faulty winding. A bad motor winding can be easily checked with an ohmmeter. The windings could be shorted, open, or grounded. The

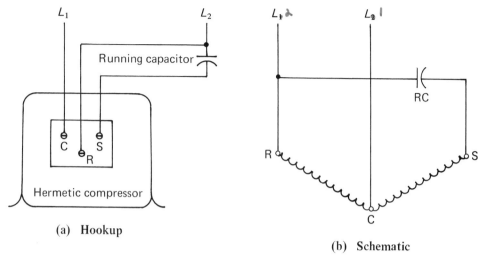

(a) Hookup

(b) Schematic

Legend

C: Common Terminal
R: Running Winding Terminal
S: Starting Winding Terminal
RC: Running Capacitor

FIGURE 6.24.   Permanent split-capacitor motor (a) Hookup (b) Schematic diagram of the hookup.

service mechanic should use care in diagnosing problems with the windings of PSC motors because they are often built with several speeds.

A bad capacitor can keep a PSC motor from starting or can pull a high ampere draw when running. Capacitors can be checked by one of the methods covered in section 6.6. In most cases faulty PSC motors will be replaced with new motors rather than repaired. The PSC motor is easy to troubleshoot with the right tools and knowledge.

Probably the most difficult aspect of PSC motors is their design. PSC motors are often built with several speeds. Figure 6.25 shows a schematic diagram of a PSC motor with three speeds. Service mechanics must pay careful attention when replacing a faulty PSC motor because if the motor is connected incorrectly, permanent damage can occur. Most PSC motors are furnished with a wiring diagram to ensure correct installation. Motor manu-

FIGURE 6.25.   Schematic diagram of a three-speed PSC motor.

FIGURE 6.26. *Capacitor-start motor. Robbins & Myers, Electric Motor Div.*

facturers, however, make only a limited number of motors to replace the many different motors in the field. Thus a service mechanic may have to adapt the replacement motor to a specific application.

6.8 CAPACITOR-START MOTORS

The **capacitor-start motor**, shown in figure 6.26, produces a high starting torque, which is needed for many applications in the industry. The open capacitor-start motor operates basically like a split-phase motor (see section 6.5) except that a capacitor is inserted in series with the centrifugal switch and the starting winding. The centrifugal switch breaks the flow of current to the starting capacitor and starting winding. The centrifugal switch opens when the motor has reached a speed that is 75% of its full speed. Figure 6.27 shows a schematic diagram of the motor.

Capacitor-start motors are used on pumps, small hermetic compressors, washing machines, and some types of heavy-duty fans.

FIGURE 6.27. *Schematic diagram of an open capacitor-start motor with a centrifugal switch (relays may be used instead of a centrifugal switch).*

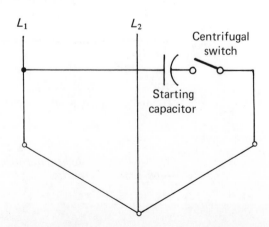

Open Type

As we have said, the open capacitor-start motor is very similar in design to the split-phase motor with the exception of the capacitor. Therefore, trouble-shooting the open capacitor-start motor is similar to checking the split-phase motor except for checking the capacitor. There are four possible areas of trouble in an open capacitor-start motor: windings, bearings, centrifugal switch, and capacitor.

The windings can be easily checked with the use of an ohmmeter by checking for shorts, opens, and grounds.

The bearings of a motor usually fail because of lack of maintenance or wear. Motor bearings will usually become tight or lock down completely. This condition can be determined by trying to turn the motor. If the motor has a tight place in its rotation or will not turn at all, the bearings are faulty in the motor.

Due to the constant opening and closing of the centrifugal switch, it is often the culprit in motor problems. The centrifugal switch may stick in an open or closed position. Or its contacts may be defective. A centrifugal switch in some cases can be checked with an ohmmeter to determine its position, open or closed. In other cases the motor will have to be disassembled to check the switch.

The capacitor is easy to check with an ohmmeter. The capacitor is often mounted in one end bell of the motor rather than on the top of the stator.

Enclosed Type

When capacitor-start motors are used in small hermetic compressors, a centrifugal switch cannot be used because of the oil that is used to lubricate the compressor. Instead an external relay is used to break the power going to the starting winding and the starting capacitor. In this case the capacitor-start motor is an enclosed motor with a starting relay. By inserting a capacitor in the starting winding, a phase displacement is created between the running and starting windings, causing the motor to rotate.

The enclosed capacitor-start motor has an external relay to drop the starting capacitor out of the circuit. This capacitor should be checked to determine its condition.

The condition of the windings of an enclosed motor can easily be checked with an ohmmeter. The windings have a set of terminals on the outside of the casing that lead to the windings. Use an ohmmeter to check across these terminals to determine if the windings are shorted, open, or grounded.

The enclosed motor can also be locked down due to the bearings or to the internal failure of some component of the motor. This condition can be detected with an ammeter or by the humming sound of the motor on an attempted start.

6.9 CAPACITOR-START–CAPACITOR-RUN MOTORS

The capacitor-start–capacitor-run motor (CSR), produces a high starting torque and increases the running efficiency. It is actually a capacitor-start motor with a running capacitor added permanently to the start winding. The starting winding is energized all the time the motor is running. The capacitor-start–capacitor-run motor takes the good running characteristics of a permanent split-capacitor motor (see section 6.7) and adds the capacitor-start feature. This produces one of the best all-round motors used in the industry.

Capacitor-start–capacitor-run motors are used almost exclusively on hermetic or semihermetic compressors. Rarely will this type of motor be used as an open-type motor because of the cost of the components necessary to produce it. Most open-type motors do not use a starting relay but use the centrifugal switch instead. Open types of motors are usually built as a permanent split-capacitor motor or a capacitor-start motor. Occasionally a CSR motor will be used in an open-type motor when an extremely high starting torque is required.

Operation

The CSR motor begins operation on a phase displacement between the starting and running windings, which allows rotation to begin. The running capacitor lends a small amount of assistance to the starting of the motor, but its main function is to increase the running efficiency of the motor. Figure 6.28 shows a schematic diagram of this motor.

Troubleshooting

The capacitor-start–capacitor-run motor is sometimes difficult to troubleshoot because of the number of components that must be added to a regular motor to produce it. The windings, bearings, potential relays, starting capacitor, and running capacitor must all be checked.

The windings of a CSR motor can be easily checked with an ohmmeter to determine if the windings are shorted, open, or grounded. In most cases the windings will be enclosed in a hermetic casing and the terminals will be on

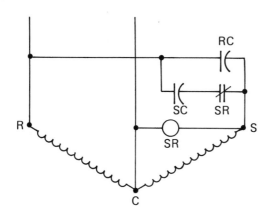

FIGURE 6.28. Schematic diagram of a capacitor-start–capacitor-run motor.

the outside of the casing. However, the type of motor makes little difference in checking the winding as long as the mechanic uses the correct terminals.

The bearings of a CSR motor can be worn so badly that the motor will not turn or will turn only with a great deal of difficulty. The bearings of hermetically sealed motors are enclosed and therefore harder to check, but the conditions of the bearings can be determined by a whining sound, or by the motor pulling a larger-than-normal ampere draw. Care should be taken not to condemn the bearings of a motor because of a high ampere draw unless you are sure that this is the problem.

The starting relay can be checked by diagnosing the condition of the contacts and the coil. The contacts can be checked with an ohmmeter or by visual inspection. On an ohmmeter the contacts should show zero resistance. The visual inspection is easy once the relay is disassembled. Then the condition of the contacts can be determined: sticking, pitting, or misalignment. The coil is checked just like the windings of a motor.

The starting and running capacitors are easily checked with an ohmmeter to determine their condition.

Troubleshooting a CSR motor is done by checking all components of the motor. These motors must be correctly checked to prevent other components from being destroyed. For example, a capacitor will be destroyed if the contacts or coil of a starting relay are bad.

6.10 THREE-PHASE MOTORS

Three-phase motors are rugged and reliable and more dependable than other types of motors. The most common type, and the type used most often in

FIGURE 6.29.   Three-phase squirrel cage motor. Photo courtesy of Reliance Electric Company.

heating, cooling, and refrigeration, is the squirrel cage induction type, shown in figure 6.29. This motor will be the only three-phase motor discussed in this chapter.

Three-phase motors are considerably stronger than single-phase motors because of the three phases that are fed to the motor. Three-phase current actually gives three hot legs to the device, rather than the two hot legs that single-phase supplies. Therefore, instead of having a two-phase displacement, a three-phase displacement is available without using starting components. Three-phase motors are common to the industry and thus the technician should understand their operation.

Operation

Three-phase motors operate on the same principles as the single-phase with the exception of the three-phase displacement. A rotating magnetic field is produced in the stator. This interacts and causes a magnetic field in the rotor. However, the three-phase motors require no starting apparatus, because none of the phases are together. In the sine wave of the three-phase motor, none of the phases peaks at the same time. Each phase is approximately 60 electrical degrees out of phase with the others. For this reason there is no need to use any device to cause a phase displacement, as is needed in the starting of single-phase motors.

Three-phase motors can be purchased in almost any voltage range desired. For example, a dual-voltage, three-phase motor can be operated on two different voltages with minor modifications in the wiring.

Three-phase motors have two basic types of windings. They are the **star winding**, as shown in figure 6.30, and the **delta winding**, as shown in figure 6.31. There is no operational difference between the two types, but it does allow designers more latitude in three-phase-motor design.

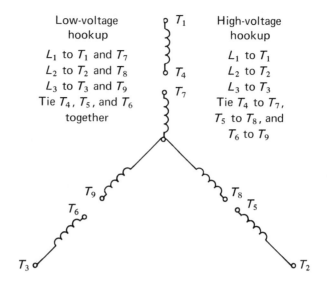

Low-voltage
hookup

L_1 to T_1 and T_7
L_2 to T_2 and T_8
L_3 to T_3 and T_9
Tie T_4, T_5, and T_6
together

High-voltage
hookup

L_1 to T_1
L_2 to T_2
L_3 to T_3
Tie T_4 to T_7,
T_5 to T_8, and
T_6 to T_9

FIGURE 6.30.   *Schematic diagram of the star winding of a three-phase motor.*

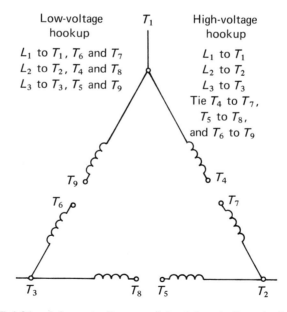

Low-voltage
hookup

L_1 to T_1, T_6 and T_7
L_2 to T_2, T_4 and T_8
L_3 to T_3, T_5 and T_9

High-voltage
hookup

L_1 to T_1
L_2 to T_2
L_3 to T_3
Tie T_4 to T_7,
T_5 to T_8,
and T_6 to T_9

FIGURE 6.31.   *Schematic diagram of the delta winding of a three-phase motor.*

143

Troubleshooting

A three-phase motor can be checked by reading the resistance of the winding with an ohmmeter. If a resistance reading of 0 ohm occurs, the motor is shorted. A reading of infinite resistance indicates an open winding. A reading of some measurable resistance is usually from 1 ohm to 50 ohms, depending on the size of the motor. The larger the motor, the smaller the resistance. The smaller the motor, the larger the resistance of the winding. Care should be taken because of the chance of a spot burnout in the winding. Experience should give service personnel the ability to diagnose any type of electric motor.

6.11   HERMETIC COMPRESSOR MOTORS

Hermetic compressors are becoming increasingly popular because of their low cost. Hermetic motors are of the induction type. They are designed for single- and three-phase current. There are four basic types of single-phase motors used in hermetic compressors. The split-phase is used on small equipment (fractional horsepower). The capacitor-start is also used on small equipment. The permanent split-capacitor is used on most window units and small residential units. The capacitor-start–capacitor-run is used on any application that requires a good starting and running torque. Many hermetic compressors are built with three-phase motors; usually these are used on the larger equipment.

Operation

Hermetic compressor motors are totally enclosed within a shell with refrigerant and oil. Hence they require special considerations. Nothing can be used inside the shell that is capable of causing a spark or that has to move on the crankshaft, such as a centrifugal switch. Therefore, no starting apparatus can be incorporated inside the compressor shell. Starting relays and capacitors must be mounted and wired outside the motor. It must be remembered that hermetic motors operate the same as other motors with the exception of the enclosure.

Terminals and Troubleshooting

All single-phase motors have a common, a start, and a run terminal. These terminals are sometimes wired directly into an open-type motor and are dif-

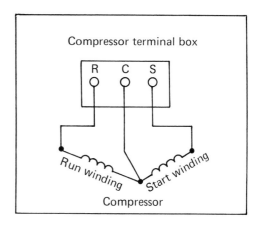

FIGURE 6.32. Schematic diagram of a single-phase compressor with the terminals identified.

Legend

R: Run Terminal
C: Common Terminal
S: Start Terminal

ficult to find. The common is the junction point of the start and run terminals. The start and run terminals are connected to one end of the windings while the common is connected to the other end. The schematic diagram of a single-phase compressor motor is shown in figure 6.32 with the terminals identified. Of course, each of the windings of a three-phase hermetic motor is the same because no starting apparatus is required.

In single-phase motors it is important for the service technician to determine the common, start, and run terminals. This task can be performed simply and easily by using an ohmmeter to obtain the resistance of each winding with respect to common. Figure 6.33 shows the ohmic values of a single-phase motor after the resistance has been measured at each terminal on the compressor. To find the run, start, and common terminals, the following procedure should be followed:

1. Find the largest reading between any two terminals. The re-

FIGURE 6.33. Terminals of a single-phase compressor with ohmic values given.

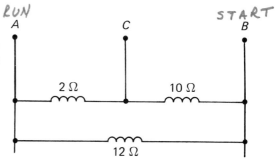

maining terminal is common (in figure 6.33 the reading between A and B is largest; C is common).

2. The largest reading between common and the other two terminals is start (C to A is 2 ohms and C to B is 10 ohms; therefore, common to B is larger and B is start).

3. The remaining terminal is run (A is run).

This procedure is important, especially in installing the external electric devices, although it is not necessary if a good readable diagram is available. In a three-phase motor the resistances between all three terminals are the same.

SUMMARY

The single most important operating principle of an electric motor is the rotating magnetic field produced by alternating current. An alternating current is applied to the stator of the motor to produce a magnetic field. The magnetic field interacts with the rotor to produce a magnetic field in it. When these two magnetic fields act together, they produce a rotating movement in the motor.

There are basically five types of motors used in the industry: shaded-pole, split-phase, permanent split-capacitor, capacitor-start–capacitor-run and capacitor-start, and three-phase.

The shaded-pole motor is a low-starting-torque motor that is used on some propeller types of fans. They are easy to identify because of the copper band around the shaded pole. They are easily diagnosed for trouble because of their simple winding patterns.

Split-phase motors are relatively low torque motors. They are simple and inexpensive devices. A split-phase motor can be used on a small hermetic compressor with the starting winding being dropped out by a starting relay. Or it can be used with any open types of motors that do not require a high starting torque.

The permanent split-capacitor motor has a low starting torque and a running capacitor in the starting winding. The running capacitor remains in the circuit at all times to produce good running efficiency. The permanent split-capacitor motor is used on most residential air conditioners of 5 horsepower or under and on direct-drive fan motors. It is relatively inexpensive because it does not have a switch to drop the starting winding.

The capacitor-start and the capacitor-start–capacitor-run motors are similar in design. They have a high starting torque. The addition of a running capacitor to a capacitor-start motor produces the capacitor-start–capacitor-run motor, which has good running efficiency.

Three-phase motors are commonly used on large pieces of equipment. They operate much like single-phase motors except that they have three basic phase displacements without the use of any starting apparatus. They have better starting and running characteristics than single-phase motors.

Hermetic compressor motors are becoming popular because of their low cost. They are used in many cooling units, especially for the smaller systems. Hermetic motors must have all their starting apparatus wired and mounted externally because it cannot be contained in the compressor shell. A hermetic compressor motor may be any one of the five basic motor types discussed in this chapter. The enclosure is what distinguishes the hermetic compressor from the other motor types.

Almost all single-phase motors require some method of producing a second phase of electricity in the motor to make it start. The design of the windings in a split-phase motor allow the use of a centrifugal switch to drop the starting winding out of the circuit after the motor has attained 75% of its full speed. Other types of single-phase motors use capacitors to create the second phase. The permanent split-capacitor motor incorporates a running capacitor to aid in the starting of the motor. The capacitor-start–capacitor-run motor incorporates the running capacitor along with a starting capacitor, using a potential relay to drop out the starting capacitor. The capacitor-start motor operates as a split-phase motor except that a start capacitor is added to the start winding.

QUESTIONS

1. What is magnetism?

2. What is torque?

3. What is a magnetic field?

4. What material conducts a magnetic field best?

5. True or false. Unlike poles of a magnet repel each other and like poles attract.

6. What part does polarity play in the operation of an electric motor?

7. What part does the frequency of alternating current play in the operation of an electric motor?

8. The electromagnet is a magnet produced by _____.

9. What is a squirrel cage rotor?

10. What are the five types of single-phase motors used in the industry?

11. A rotor is _____ and a stator is _____.

12. List the five types of motors according to their starting strength.

13. How does a shaded-pole motor operate?

14. How can a shaded-pole motor be reversed?

15. How does the split-phase motor produce the phase displacement necessary to begin rotation?

16. Split-phase motors are used when a _____ torque is not required.

17. What is the purpose of a centrifugal switch in a split-phase motor?

18. True or false. Split-phase motors are the most reliable motors used in the industry.

19. What is the purpose of a capacitor?

20. What are the two types of capacitors and their purpose?

21. The unit of measurement for capacitors is the _____.

22. List the five capacitor replacement rules.

23. What are the procedures to use in checking capacitors?

24. Explain the operation of a permanent split-capacitor motor.

25. The phase displacement in the PSC motor is created by _____.

26. What are the advantages of using a capacitor-start–capacitor-run motor?

27. The major difference between a split-phase motor and a capacitor-start motor is that _____.

28. What is the purpose of using a capacitor in series with the starting winding if the winding is in parallel with the running winding?

29. True or false. CSR motors are used almost exclusively on hermetic compressors.

30. Why is a starting apparatus unnecessary when using a three-phase motor?

31. What are the advantages of using a three-phase motor over a single-phase motor?

32. Why must all starting apparatus be mounted externally to a hermetic compressor?

33. How can the common, start, and run terminals of a single-phase motor be determined?

34. True or false. Capacitor-start motors are built with several different speeds.

7

Components for Electric Motors

INTRODUCTION

In the preceding chapter we discussed electric motors and some of their starting components. In this chapter we will discuss starting components used on hermetic motors. Hermetic and other special motors require this external starting component because they are enclosed in a sealed case. There are three types of starting relays that are used on this type of motor: current, potential, and hot-wire relays. These devices are used on most hermetic compressor motors with the exception of permanent split-capacitor motors.

Electric motors must have some type of bearings to allow for smooth and easy rotation. The ball bearing or the sleeve bearing are used in almost all motors. Motors also must have some means of transferring their rotating motion to the device that is being powered by the motor. A direct-drive hookup transfers the rotating motion directly from the motor to the device. The belt-drive hookup transfers the rotating motion to the device by a belt connection.

Magnetic starters and push buttons are used to stop and start electric motors. A magnetic starter opens and closes sets of contacts to stop and start loads. The magnetic starter also incorporates overload protection for the device it controls. Push-button switches are switches that are used to control magnetic starters in most cases.

7.1 STARTING RELAYS FOR SINGLE-PHASE MOTORS

Single-phase motors, with the exception of the permanent split-capacitor motor, must have some means of dropping the starting winding (or the starting capacitor in the case of a capacitor-start–capacitor-run motor) out of the circuit. In an open-type motor this is accomplished simply and easily by a centrifugal switch mounted in the motor. The switch opens the starting circuit once the motor reaches 75% of its full speed. In enclosed motors some type of **starting relay** must be used.

There are basically three types of starting relays used in the industry today. The current type and the hot-wire relay are generally used on small hermetic motors. The potential relay can be adapted to any size unit. Starting relays are essential to the operation of a sealed motor compressor.

Each of the three types of starting relays uses a different method to drop the starting circuit in or out. A potential relay operates on the principle of **back electromotive force.** The back electromotive force is the amount of voltage produced in the starting winding of a motor. The current or magnetic type of relay operates on the current or amperage that the motor uses to start. Hot-wire relays use current flow to produce heat across a thermal element, which operates the starting circuit. Each of these relays must be correctly sized and matched with the application. Each relay is designed to remove the starting circuit when the motor reaches approximately ⅔ or ¾ full speed.

7.2 CURRENT OR AMPERAGE RELAYS

On all motors the starting amperage is greater than the running amperage because the rotor is at a standstill on start-up. Therefore, the ampere draw of the motor is high at the time of the initial start-up. But as the motor gains speed, the ampere draw decreases. The **current or magnetic relay** uses the electrical characteristics of the motor to remove the starting circuit electrically once the motor has established a good running speed.

Operation

The current relay is built much like a solenoid, as shown in figure 7.1. The contacts of the current relay are normally open. As the motor attempts to start, the amperage increases. This makes the magnetic field of the relay stronger, closing the contacts.

FIGURE 7.1. Current relay. Photos courtesy of General Electric Co., Appliance Control Dept. and Texas Instruments Incorporated.

Figure 7.2 shows the position of the contacts when the motor is in the starting phase. As the speed of the motor increases, the amperage decreases. When the motor has reached ⅔ or ¾ of its full speed, the amperage will be low enough to cause the magnetic field strength of the relay coil to decrease enough to drop the relay contacts out of the starting circuit.

Troubleshooting

Most current relays are easy to troubleshoot because they have a coil and a set of contacts that can be easily checked with an ohmmeter. The current relay has normally open contacts that are easily checked by turning the relay upside down and checking the contacts with an ohmmeter. If the contacts

FIGURE 7.2. Schematic diagram of a current relay when it is energized.

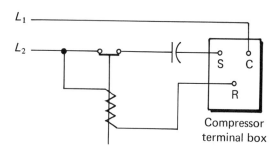

are good, the relay contacts will read open when the relay is right side up. They will read closed when the relay is inverted.

The coil of the relay is made of large wire and should have a very low resistance, around 0 to 1 ohm. If the coil reads any higher, more than likely the coil is bad and the relay should be replaced. The current relay and motor starting components should be completely checked to ensure that no other components are bad.

7.3 POTENTIAL RELAYS

The **potential relay** is gaining popularity because it can be adapted easily to almost any compressor. The back electromotive force produced by the starting winding of a motor is the controlling factor of a potential relay.

Operation

When a single-phase motor is operating, there is a voltage produced across the starting windings above and beyond the voltage being applied to the motor. The starting windings actually act as a generator to produce the back electromotive force of a motor. The back electromotive force of a motor corresponds to the motor speed. The potential relay is designed to open, dropping the starting circuit, when the motor reaches a certain back electromotive force that is predetermined by the manufacturer of the motor.

FIGURE 7.3. Potential relay. Photos courtesy of General Electric Co., Applicance Control Dept. and Essex Group, Controls Div.

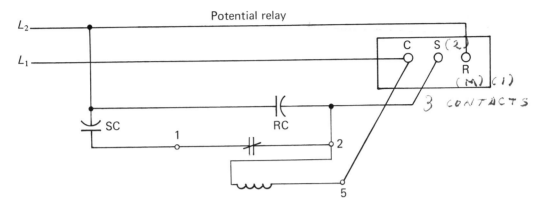

FIGURE 7.4.   Schematic diagram of a potential relay.

Figure 7.3 shows several potential relays that looks much like an ordinary general-purpose relay. The wiring diagram for a potential relay is shown in figure 7.4. The potential relay has an advantage over other relays because its contacts are normally closed when the unit starts and causes no arcing. Arcing is an electric spark that is produced across two sets of contacts. As the motor speed increases, so does the back electromotive force. When the speed approaches 75% to 80% of full speed, the back electromotive force will be large enough to drop the starting circuit by energizing the potential relay coil.

Troubleshooting

The potential relay is easy to troubleshoot because there are only two parts of the relay that must be checked. The coil of the potential relay can be checked with an ohmmeter across 2 and 5 in figure 7.4, which could check out as good, open, or shorted. The contacts of the potential relay across 1 and 2 should be checked, and they should be closed. Motor current will increase if the potential relay does not drop the starting circuit at the proper time. This condition is dangerous to the motor and the starting apparatus. The potential relay should always be checked completely to prevent this condition from occurring. A service technician should always check the operation of a unit after the dropout of a starting relay. The correct size relay should always be used for replacement of a faulty relay. If the starting relay is working properly, there will be no ampere draw through the starting circuit after the motor has reached full speed.

7.4 HOT-WIRE RELAYS

The **hot-wire relay**, shown in figure 7.5, operates on the principle that electric energy can be converted to heat. The relay uses two bimetal strips. One strip operates the starting circuit; a second strip acts as an overload for the running winding.

Operation

Figure 7.6 shows the schematic of a hot-wire relay. From L to A is the actual hot wire, which would heat up the bimetal elements B and C. C goes to the start winding and B to the running winding. If the hot wire reaches a temperature high enough to cause element C to warp, the starting circuit is dropped out. This is the correct procedure. If the hot wire reaches an even higher temperature, this causes element B to warp and open the circuit to the running winding. Then the compressor would cut off on overload by the relay. Some manufacturers allow the hot wire to stretch as it gets hot to open and close the contacts. Both types (warping and stretching) are similar in operation. The schematic in figure 7.6 shows a hot-wire relay hooked up in a system.

Troubleshooting

The hot-wire relay is probably the hardest starting relay to check. The relay is hard to check because it is hard to detect completely the heat that is being supplied to the relay. Even the slightest temperature difference will cause the relay to show an overload.

The contacts of the relay are checked with an ohmmeter. The diagnosis

FIGURE 7.5. Hot-wire relay. Delco Products Division, General Motors Corp.

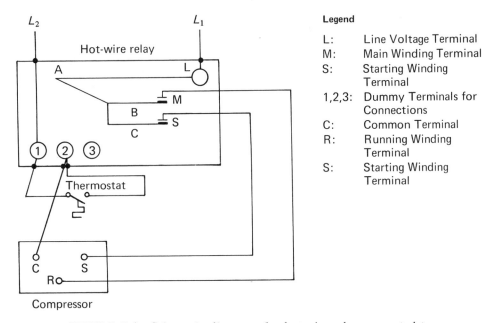

Legend

L:	Line Voltage Terminal
M:	Main Winding Terminal
S:	Starting Winding Terminal
1,2,3:	Dummy Terminals for Connections
C:	Common Terminal
R:	Running Winding Terminal
S:	Starting Winding Terminal

FIGURE 7.6. Schematic diagram of a hot-wire relay connected to a compressor.

of the hot wire and thermal elements of the relay is done while the unit is operating or starting. The first thing to check for is to see if the motor is running. If the motor is running but cutting out on overload, the bearings inside the motor might be tight.

The starting winding should drop in and out quickly and is, therefore, hard to detect. The amperage drawn by the starting winding should be checked to determine if it is too high. Usually the starting winding will draw a higher amperage than the running winding. As for other starting relays, in troubleshooting the hot-wire relay, the relay and all other motor components should be checked.

7.5 MOTOR BEARINGS

All rotating electric equipment has some type of bearing to allow for smooth and easy rotation. A **bearing** is that part of a rotating electric device that allows free movement. The two types of bearings used in the heating, cooling, and refrigeration industry are **ball bearings** and **sleeve bearings**, shown in figure 7.7. The ball bearing is the most efficient because it produces less friction. The sleeve bearing is cheaper, however, and is more commonly used.

FIGURE 7.7. Sleeve and ball bearings used in the industry.

Ball Bearings

Ball bearings are designed with an inner and an outer ring, which enclose the balls by use of a separator. The inner ring is the bore in which the shaft is pressed.

Ball bearings are lubricated in three ways. The permanently lubricated bearings are sealed at the factory and only rarely require additional lubrication. The packed lubrication of ball bearings is done by hand. The bearing must be disassembled and hand packed with grease every two to five years. Many ball bearings are equipped with a grease fitting and should be lubricated every two years. Lubrication is of prime importance to a motor. And note that overlubrication is as damaging as underlubrication. Figure 7.8 shows a ball bearing pressed into an end bell of a motor.

Ball bearings are used on most heavy loads. However, ball bearings cannot be used in a hermetic compressor because of the danger of sparks. Some of the many advantages of using ball bearings are mountings in any position, antifriction design, versatility, and the ability to carry a large load. Ball

FIGURE 7.8. Ball bearing inserted into the end bell of a motor.

bearings are often permanently lubricated from the factory or require infrequent lubrication. Thus they require less maintenance than a sleeve bearing. The life of a ball bearing is usually long and trouble free, but misuse can rapidly destroy a ball bearing.

Sleeve Bearings

Sleeve bearings are brass or bronze cylinders in which a shaft rotates. The bearings have more friction than the ball bearing and thus are used for lighter duty. Lubrication of sleeve bearings is accomplished by oil wick lubrication, yarn-packed lubrication, or oil ring lubrication.

The oil wick lubrication has a wick that extends into an oil reservoir. The wick picks up oil from the reservoir and transfers it to the shaft, as shown in figure 7.9. The reservoir should be filled with oil twice a year.

The yarn-packed lubrication is merely yarn packed around the shaft to lubricate it, as shown in figure 7.10. This bearing should be lubricated every few months.

The oil ring lubrication is used on large motors. In this device a ring rotates through an oil reservoir, picking up oil and transferring it to the shaft, as shown in figure 7.11. The oil level of the reservoir should be checked monthly.

Hermetic compressors use sleeve bearings because with sleeve bearings there is no danger of sparks. Care must be taken in mounting motors equipped with sleeve bearings because of the lubrication problems when the bearings are mounted in a vertical position. Sleeve bearings are commonly used in many applications in the industry.

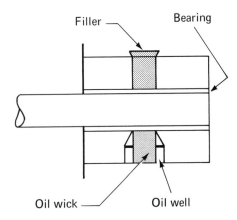

FIGURE 7.9. Oil wick lubrication of a sleeve bearing.

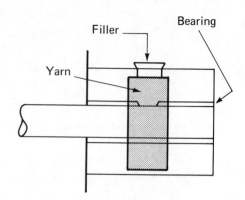

FIGURE 7.10. *Yarn-packed lubrication*
of a sleeve bearing.

7.6 MOTOR DRIVES

A motor drive is the connection between an electric motor and a component
that requires rotation. Electric motors are used to drive almost all devices that
require rotation. There are two basic driving devices: direct drive and belt
and pully drive.

Direct Drive

Direct-drive methods require that a device turn with the same revolutions per
minute as the motor. Fan motors and pumps are often direct drive; hermetic
compressors are always direct drive. Direct-drive hookups require a very
close fit between the motor and the device. In a hermetic compressor the
crankshaft is made in one piece, with the motor on one end and the compressor portion on the opposite end. Direct-drive applications usually require
a coupling, except with hermetic compressors and fan motors.

The two types of direct-drive couplings used in the industry are the
flexible-hose coupling, as used on oil burners, and flange couplings, as used

FIGURE 7.11. *Oil ring lubrication of a*
sleeve bearing.

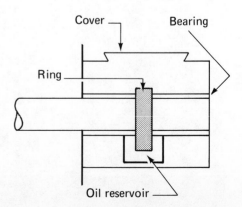

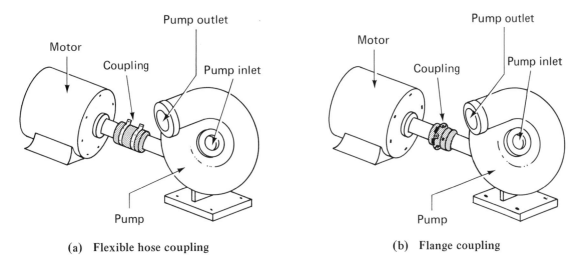

(a) Flexible hose coupling   (b) Flange coupling

FIGURE 7.12.   Two widely used direct-drive couplings.

on some open types of compressors and hot-water pumps. Figure 7.12 shows both commonly used types of direct-drive couplings.

V-Belt

Although there are many types of belts used to drive devices, only the **V-belt** will be discussed here. It is used almost exclusively in the industry. By using two pulleys and connecting them with a V-belt, rotation can be transferred from the motor to the device, as shown in figure 7.13. V-belts are used in many applications in the industry, such as driving open types of compressors, fan motors, and pumps.

There are basically three sizes of V-belts used presently. Size FHP V-belts are ⅜ inch wide and used on fractional-horsepower motors. Size A-section is ½ inch wide and used for most jobs requiring 1-to-5-horsepower motors. Size B-section is $^{21}/_{32}$ inch wide and used when the motor pulley is larger than 5 inches in diameter. More than one V-belt is often used to pull large devices. When belt changes are needed, a matched set should be used in a dual-belt application. The equipment's revolutions per minute can be changed by changing the pulley size or sizes. The following formula can be used:

$$\text{diameter of equipment pulley} = \frac{\text{motor rpm} \times \text{motor pulley size}}{\text{rpm of equipment}}$$

V-belts should always be properly aligned and the tension adjusted. Figure 7.14(a) shows an incorrectly aligned V-belt. The proper amount of tension

159

FIGURE 7.13. V-belt connection between motor and fan. Reproduced by permission of Carrier Corporation. © 1977 Carrier Corporation.

for most belts is shown in figure 7.14(b). Frequently a V-belt will slip, causing erratic equipment speed or break. In these cases the belt should be replaced and the tension adjusted correctly.

7.7 MAGNETIC STARTERS

A **magnetic starter**, shown in figure 7.15, is composed of four sets of contacts, a magnetic coil to close the contacts when the coil is energized, and a set of overloads. The coil is the heart of the system. All functions of the total control, and stopping and starting the equipment, are accomplished by the coil being energized or deenergized. The contacts open and close depending on the action of the coil. The magnetic starter also has a means of overload protection, which deenergizes the coil in the event that an overload occurs. Figure 7.16 shows a simple wiring diagram of a starter controlled by a thermostat.

In the heating, cooling, and refrigeration industry, many of the motors used are protected by internal overloads in the motor or by some external means, such as an overload relay, magnetic overload, or thermal elements. A contactor is a device that opens and closes automatically and allows voltage to flow to the equipment (refer back to chapter 3). Most air-conditioning

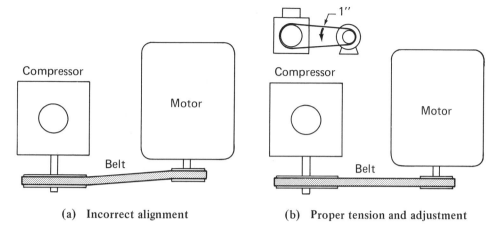

(a) Incorrect alignment   (b) Proper tension and adjustment

FIGURE 7.14.   V-belt alignments (a) Incorrect (b) Correct.

equipment uses a contactor with a separate means of overload protection. In many cases, especially with three-phase equipment, manufacturers use a magnetic starter, which incorporates a means of overload protection along with the ability to stop and start current flow to the equipment.

Types of Magnetic Starters and Their Operation

The overloads used in a magnetic starter are of three general classes: bimetal relay, thermal relay, and molten-alloy relay. The **bimetal relay** is composed of two metals welded together with different expansion qualities. One end is fastened securely and the opposite end can move. When this element is heated

FIGURE 7.15.   Magnetic starter. Photo courtesy of Furnas Electric Co.

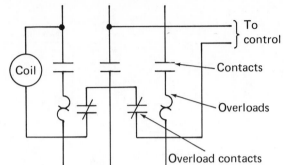

FIGURE 7.16. Schematic diagram
of a magnetic starter.

above a certain temperature due to the current flow to the equipment, it
will break, causing the starter to open.

The **thermal relay** operates on the heat going through a wire to determine
the current flow. If the heat is great enough to indicate an overload, the contacts of the relay shown in figure 7.17 will open.

The **molten-alloy relay**, shown in figure 7.18, is nothing more than a link
of material that is a good electrical conductor. A ratchet wheel is soldered
to the conductor with a special alloy. When the current flow produces a
higher-than-normal temperature, the solder melts, allowing the ratchet wheel
to open the contacts.

No matter what type of overload relay is used, it should be sized correctly
to work properly. Manufacturers of magnetic starters distribute much information on how to size starters and overloads (sometimes referred to as
"heaters").

FIGURE   7.17.   Thermal   relay.   Photo
courtesy of Furnas Electric Co.

FIGURE 7.18. Molten-alloy relay. Square D. Co.

Troubleshooting

Magnetic starters are used on most three-phase equipment because they contain overloads and can effectively control the operation of loads in a heating, cooling, or refrigeration system. Some of the common faults of magnetic starters occur in the contacts, solenoid coil heaters or overloads, and the mechanical linkage. No matter what part of the magnetic starter is faulty, it must be treated as a unit at the beginning of the diagnosis.

The service mechanic must first be certain that the magnetic starter is at fault. If the magnetic starter is at fault, it must be replaced, or the faulty section must be repaired. The contacts of a starter can be checked by visual inspection or by checking the voltage drop across each set of contacts. If the contacts are badly pitted, or if there is a voltage drop across the contacts, then they must be replaced.

The coil of a contactor can easily be faulty; it can be shorted, open, or grounded. The linkage that connects the movable contacts to the solenoid plunger can be worn or broken and will not close the contacts even if the coil is energized.

The overloads are difficult to troubleshoot because of the different types. Overloads are designed to open a control circuit when an overload exists. The overload circuit of a magnetic starter is connected within the control enclosure. The service mechanic must be familiar with the types of overloads and their operation. The size of the overload should also be checked. The mechanic can easily check a magnetic starter step by step after it has been determined that the starter is not closing and should be.

7.8 PUSH-BUTTON STATIONS

Push-button stations or switches are switches that are controlled manually by the pressing of a button. They can have two switches (for stop and start) or many switches (for such functions as on, off, start, stop, jog, reverse, and forward). They are designed in many forms and with many different functions. There are relatively few instances in the heating, cooling, and refrigeration industry where push-button stations are used. They are used mainly for motor control, circuit control, and magnetic starters.

Push-button stations are very easy to troubleshoot. The service mechanic must know, or be able to find out, the normal position of the switch. In some cases push-button switches may be quite complex, but this will not occur very often. An ohmmeter can be used to diagnose the condition of the switch.

SUMMARY

Starting relays are essential to the correct operation of an enclosed motor. They drop the starting winding or capacitor out of the circuit when the motor has reached $\frac{2}{3}$ or $\frac{3}{4}$ of its full speed.

There are three types of relays in common use today. The current relay operates on the principle that amperage flowing through a wire will produce a magnetic field. The potential relay uses the back electromotive force of a motor to drop the starting circuit. The hot-wire relay uses heat to deenergize the starting circuit to the motor. The potential relay can be used on almost any application; the current and hot-wire relays are used mainly on small equipment.

All rotating heating, cooling, and refrigeration equipment must have bearings to operate efficiently and smoothly. Ball bearings and sleeve bearings are in common use today for this purpose. Correct lubrication is essential to ensure long life and efficient service from the bearings.

In almost all cases in the industry, electric motors drive some type of equipment. Direct-drive applications are common, as are V-belt applications. Direct-drive applications require a very accurate fit with little vibration. The device being driven must turn at the same revolutions per minute as the motor. V-belt connections have a certain amount of tolerance, but they must be correctly adjusted. They can be used to alter the revolutions per minute of the equipment by changing the pulleys.

Magnetic starters allow the versatility of using one device that includes the control of the equipment along with an overload protection. They are used on large three-phase equipment and have some separate motor applications. Care must be taken when sizing overloads so that the correct size of magnetic starter is chosen.

QUESTIONS

1. What is the purpose of starting relays?

2. What is a potential relay and why is it used?

3. What is a current relay and why is it used?

4. What is the purpose of a hot-wire relay? How does it work?

5. The contacts of the current relay are _____.

6. The controlling factor of the potential relay is _____.

7. True or false. The contacts of the potential relay are normally open.

8. What are the two types of bearings used in the industry? What are the advantages and disadvantages of each?

9. What is a direct drive? How is it used in the industry?

10. Why are V-belts popular in the industry?

11. What is the correct tension on a V-belt?

12. What is arcing?

13. True or false. If the starting relay is working properly, there will be no ampere draw through the starting circuit after the motor reaches full speed.

14. Probably the hardest starting relay to check is the _____.

15. A bearing is the part of a rotating electric device that _____.

16. Name the three ways in which ball bearings may be lubricated.

17. True or false. Overlubrication of a motor and its bearings is as damaging as underlubrication.

18. What is one situation in which ball bearings cannot be used?

19. Name the three ways in which sleeve bearings may be lubricated.

20. Direct-drive hookups require a close fit between _____.

21. What is a magnetic starter?

22. True or false. The sets of contacts are the heart of the magnetic starter.

23. What are the three types of magnetic starters used in the industry? How does each device work?

24. True or false. Magnetic starters are used primarily on single-phase equipment.

25. True or false. Push-button stations are widely used in the industry.

8

Contactors, Relays, and Overloads

INTRODUCTION

Control systems used on modern heating, cooling, and refrigeration systems use many different control components to obtain automatic control. The purpose of a control system is to automatically control the temperature of some medium. The function of a control system is to stop and start electric loads that control the temperature of the medium. In the case of an air-conditioning or heating system, the primary purpose is to control the temperature within a certain area. In a refrigeration system, the purpose is to control air temperature or water temperature.

In an air-conditioning or refrigeration system, the compressor is the largest load and usually requires a contactor or magnetic starter to energize it. Loads that require more control and are too large for line voltage thermostats or manual switches use relays for the proper control. Relays and contactors work similarly. The main difference between them is their current-carrying capacity. The contactor can handle a large ampacity. Relays are usually limited in the ampacity they can carry.

In all control systems there must be a means of protecting the loads. The overload is used to protect loads such as compressors, heaters, fan motors, and pumps. The basic overload device is the common fuse. However, the fuse is inadequate to protect effectively all the important loads in a control system. Thus more effective devices for overload protection are used. As we will see, overloads come in many designs and sizes.

All electric control components serve a definite purpose in the total control system. Fortunately, many manufacturers now use simple control systems with fewer components in their residential air-conditioning control systems. However, the commercial and industrial control systems are fairly complex, with more components and better control than the residential equipment. Therefore, it is essential that heating, cooling, and refrigeration personnel become familiar with control components and understand them so that they can diagnose faulty components and perform effective control system troubleshooting.

In this chapter we will look at the components of a control system that control the loads in the system. In succeeding chapters we will discuss other control system components and the methods for troubleshooting these systems.

8.1 CONTACTORS

A contactor (see figure 8.1) is used to control an electric load in a control system. Contactors make or break a set of contacts that controls the voltage applied to some load in cooling systems. They isolate the voltage controlling its magnetic coil from the voltage applied to the load. A contactor consists of a coil that opens and closes a set of contacts due to the magnetic attraction created by the coil when it is energized. Magnetic starters are also used to start and stop large loads in cooling systems. The major difference between magnetic starters and contactors is that the magnetic starter can house its own overload. Magnetic starters were discussed in chapter 7.

Applications

The largest electric load in any cooling system that requires control is the compressor. In smaller equipment several other loads might be connected

FIGURE 8.1. Contactors. Photos courtesy of GTE Sylvania Inc., Stamford, Ct. and Essex Group, Controls Div.

in parallel with the compressor. Larger systems usually maintain a switching device for each component. The contactor used in a small residential air-conditioning unit probably controls the compressor and condenser fan motor. Figure 8.2 shows a wiring diagram of a small residential unit. Large air-conditioning units usually have several contactors. A large condensing unit, for example, might use two contactors for the compressor and three for the condenser fan motors. Large electric resistance heating systems also have several contactors. For example, they might use a contactor for each section of heaters and for some method of controlling the fan or air supply.

Operation

Different manufacturers design contactors in different ways. But all contactors accomplish the same purpose: opening and closing a set of contacts. The armature of a contactor is the portion that moves. The movement of the armature can be accomplished in basically two ways, with a sliding armature or a swinging armature. The sliding armature is shown in figure 8.3 and the swinging armature is shown in figure 8.4. The sliding armature mounts between two slots in the frame of the contactor and moves up and down in these slots. The swinging armature is mounted on a line and moves up and down in a swinging motion.

The armature of a contactor is connected to a set of contacts that causes a completed circuit when the armature is pulled into the magnetic field pro-

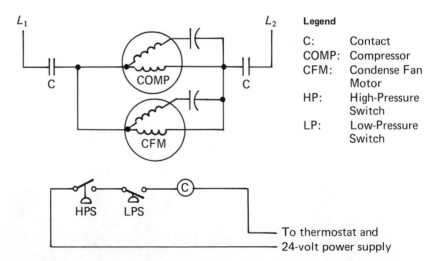

FIGURE 8.2. Schematic diagram of a small residential air-cooled condensing unit with a contactor controlling the compressor and condenser fan motor.

FIGURE 8.3. Contactor with a sliding armature. Photo courtesy of Furnas Electric Co.

duced by the coil. This operation is true for both the sliding armature and the swinging armature. The magnetic field that closes a contactor is created by a coil wound around a laminated iron core, as shown in figure 8.5. When the coil is energized, a magnetic field is created around the laminated core. The core then becomes an electromagnet of sufficient strength to attract the armature closing the contacts. Both types of contactors use the same principle of operation. Some contactors have springs mounted between the armature and the stationary contacts to ensure that the contactor opens when the coil is deenergized.

FIGURE 8.4. Contactor with a swinging armature. Photo courtesy of Essex Group, Controls Div.

FIGURE 8.5. Solenoid coil of a contactor. Photo courtesy of Essex Group, Controls Div.

Coils

Coil characteristics depend on the type of wire and the manner in which it is wound. The potential coil is energized by a certain voltage being applied to it. Coils of this type are designed to be operated on 24 volts, 110 volts, 208/230 volts, and occasionally 460 volts. The coil is identified by the voltage marked on it, as shown in figure 8.5. The potential coil is used on many special relays in the industry. The connection of a coil is usually made directly on the terminals of the coil. But in some cases the connections are jumped to a section of the contactor frame.

Contacts

The contacts of a contactor make a complete circuit when the contactor is energized, allowing voltage to flow to the controlled load. Contactors are rated by the ampere draw they can carry. There are two types of loads that a contactor can control: an inductive load, such as a motor, which has a higher ampere draw on start-up than while running; and a resistive load, which has a constant ampere draw, such as a resistance heater. Some contactors are rated for both inductive loads and resistive loads. So care should be taken when selecting a replacement contactor. The ratings of contactors are usually marked on the contactor frame.

Contacts are made of silver and cadmium, which resists sticking. The contacts are connected to a strong backing by mechanical or chemical bonding. The chemical composition of contacts is such that they operate at cool temperatures at up to 125% of their current-carrying capacity. Contactors are usually manufactured with two or three poles and in many cases four. The fourth pole is used to interlock some load device into the system or can be left unused. A two-pole contactor is required for single-phase systems. A contactor with at least three poles is required for three-phase systems.

Troubleshooting

The diagnosis of a faulty contactor encompasses three sections of the contactor: the coil, the contacts, and the mechanical linkage. A defect in any part of a contactor can cause the total contactor to be faulty.

Coil. The coil of a contactor must be in good condition to create a strong enough electromagnetic force to pull in the contactor. The coil of a contactor very rarely becomes so weak that it does not close the contacts, unless there is excessive friction of the mechanical linkage. The coil can be diagnosed as good, open, or shorted. The open and shorted conditions indicate a bad coil and can be checked with an ohmmeter. If the coil is shorted, the resistance reading will be zero. If the coil is open, the resistance will be infinite. A measurable resistance usually indicates a good coil.

A coil can also be checked by applying voltage to it and observing the contactor to see if it closes. The voltage reading of a coil should be taken before checking the coil to see if the contactor should be closed. Care should be taken when the diagnosis leads to a shorted contactor coil. If voltage is applied to it, the coil will cause a direct short and other damage could result.

Contacts. The contacts of a contactor must be in good condition to ensure that the proper voltage reaches the load. In most cases a visual inspection is sufficient to diagnose bad contacts. Figure 8.6 shows a good set and bad set of contacts for comparison.

In some cases a voltage reading taken across the contacts of the same pole will show the voltage drop across the contacts. Figure 8.7 shows the proper procedure for testing a set of contacts in this manner. The voltage in-

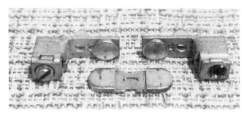

(a) Good set   (b) Bad set

FIGURE 8.6.   A good set and a bad set of contacts.

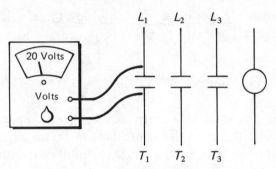

FIGURE 8.7. *Procedure to use in testing a set of contacts with a volt-meter.*

dicated on the meter is the voltage drop across the contacts (the voltage lost to the equipment). The 20 volts shown on the meter are considered to be excessive. Hence the contactor should be replaced or repaired. Any voltage above 5% of the rated voltage for the equipment is considered to be excessive. The contactor must be closed with voltage applied to make this check.

Mechanical Linkage. Probably the easiest faults to diagnose with a contactor are problems with the mechanical linkage. In most cases, any trouble with the mechanical linkage can be detected by visual inspection. Or problems can be detected by breaking the power supply and manually moving the armature of the contactor to see if the movement is free and without excessive friction. The mechanical linkage of a contactor will usually fail because of wear, corrosion, or moisture. In many cases when a contactor coil burns out, it will heat the coil and cause the varnish of the coil to gum up the contactor. For a contactor to operate properly, it must seat the contacts accurately and have free moving parts.

Repairing. Contactors can be repaired by using replacement parts from the manufacturer or a wholesaler if time permits. However, it is often difficult to find all the necessary parts, such as contacts and coils, because it is almost impossible for a wholesaler to stock all the components needed for all contactors. Most manufacturers do sell a kit that will completely replace the contact portion of the contactor. But since parts are difficult to obtain, it is usually advisable to purchase a new contactor instead of repairing the faulty one. However, be careful to choose the correct contactor for the particular application and size.

8.2 RELAYS

Relays are used to open and close a circuit to allow the automatic control of a device or circuit. Relays are similar to contactors with the exceptions of the pole configuration and the amount of current that each device can effectively handle. Relays can be used to control almost any device in the system within a certain ampacity limit.

Operation

Relays are built with the same components as a contactor. There are a coil, contacts, and some type of mechanical linkage to open and close the contacts when the relay coil is energized. When voltage (or current in some cases) is applied to a relay, it will close because of the magnetic field created in the coil and iron core. This magnetic field causes the armature of the relay to be attracted to the electromagnet created by the coil and its core. Figure 8.8 shows several different commonly used relays. The coil of a relay can be energized by voltage or current draw. The general type of relay controlling a

FIGURE 8.8.   Several different types of relays used in the industry. Photos courtesy of Essex Group, Controls Div. and Potter & Brumfield Div., AMF Inc.

device is closed by voltage. The current relay is used to control some starting device when used with a small hermetic motor.

Relays can be purchased with almost any type of pole configuration. Normally open or normally closed contacts are both utilized in control circuits. The normally open contact opens when the relay is deenergized and closes when the relay is energized. The normally closed contact closes when the relay is deenergized and opens when the relay is energized. The normal position of the relay denotes the position of any controlling device in the deenergized position. The most common types of pole configuration for relays are single-pole–single-throw, single-pole–double-throw, double-pole–single-throw, and double-pole–double throw. Any of these configurations can be normally open or normally closed.

Applications

Relays can be used to control indoor fan motors, condenser fan motors, damper motors, starting capacitors, and control lockouts. They are used for any device that requires an automatic means of opening and closing a circuit.

The indoor fan relay is a good example of the use of a relay. On the cooling cycle the indoor fan must be energized. This is accomplished by use of an indoor fan relay, as shown in the diagram in figure 8.9. The indoor fan relay will energize when the system switch and the fan switch or the cooling thermostat are closed, starting the fan motor. The indoor fan relay can also be used to control a two-speed fan motor, using high speed on cooling and

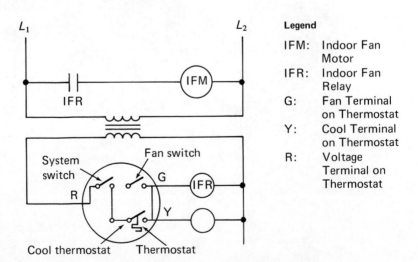

FIGURE 8.9. *Schematic diagram of an indoor fan relay circuit.*

low speed on heating with a thermal switch controlling the fan motor on heating. Relays can also be used to control contactors in a control system and for other purposes.

The potential or voltage-type relay is energized when voltage is applied to the relay coil. This relay is used to control some load device by opening and closing its contacts. The voltage-type relay can be used for many purposes, such as indoor fan relays, condenser fan relays, control relays, and lockout relays.

There are three types of relays used to assist in starting motors, as we saw in chapter 7. The potential relay uses voltage to energize its coil and drop the starting apparatus out of the circuit. The current relay uses current flow to energize the circuit that contains the starting apparatus and then drops the circuit out when the current has dropped. The thermal relay uses heat to open and close starting circuits when it is used with a motor. Starting relays for motors are used throughout the industry. However, these relays should not be confused with the general type of relay used to control loads.

Construction

The contacts used in relays are made just like contacts in a contactor. The contact is made of a silver and cadmium alloy attached to some kind of strong backing that can withstand the pressure exerted by the armature.

A relay is usually mounted in a plastic enclosure. Hence visual inspection is not as easy for relays as it is for contactors. The contacts of a relay cannot be seen unless the relay is disassembled or the cover is removed. The contacts of a relay are shown in figure 8.10.

The armature of a relay can be swinging or sliding as shown in figure 8.11.

FIGURE 8.10. Contacts of a relay. Photo courtesy of Potter & Brumfield Div., AMF Inc.

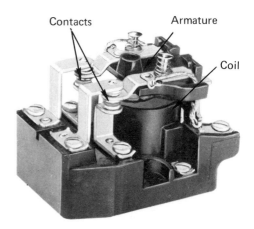

Armature

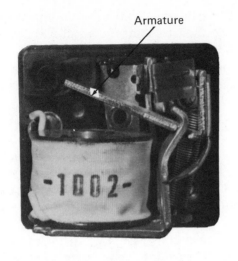

Armature

(a) Swinging armature   **(b) Sliding armature**

FIGURE 8.11.   Two types of relays used in the industry. Photos courtesy of General Electric Co., Appliance Control Dept. and Allen-Bradley Company.

These devices operate in relays in the same way that they do in contactors.

The coil of a relay is built to produce enough magnetism to effectively close the contacts of the relay. The size of the relay coil is smaller than the coil of a contactor. Figure 8.12 shows a comparison of a coil used in a relay and a coil used in a contactor.

Troubleshooting

The diagnosis of a faulty relay is done in much the same way as the diagnosis of a faulty contactor. Diagnosis of a coil is the same whether it is for a relay or for a contactor. The contacts of a relay are usually hidden and cannot be visually inspected without disassembly. They must be checked with an ohmmeter. The contacts of a relay are not as heavy as the contacts of a contactor and therefore can take less punishment. Normally open and normally closed contacts of a relay will be completely melted if a different phase of voltage is applied. This must be taken into consideration when troubleshooting contacts of relays.

The mechanical linkage of a relay gives less trouble than a contactor be-

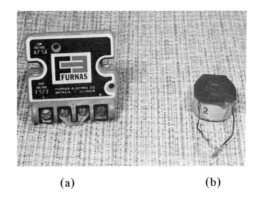

FIGURE 8.12. Comparison of solenoid coils for a contactor (a) and a relay (b). Photos courtesy of Furnas Electric Co. and Essex Group, Controls Div.

(a)   (b)

cause of the lighter weight of the armature. Any mechanical linkage problem in a relay will usually be caused by sticking contacts.

8.3 OVERLOADS

An overload is an electric device that protects a load from a high ampere draw by breaking a set of contacts. The simplest form of overload protection is the fuse. Fuses can be used to protect wires and noninductive loads, but they provide inadequate protection for inductive loads. A load that is purely resistive in nature with no coils to cause induction is called a **noninductive load**. The most common noninductive load used in the industry is an electric heater.

Fuses

Fuses consist of two ends or conductors with a piece of wire that will melt and break the circuit if the current passing through it exceeds the amperage rating of the fuse. Fuses are available in many different styles and designs (see chapter 3). Fuses are used most commonly to protect wires, circuit components, and noninductive loads. Electric resistance heaters are the most common noninductive loads protected by fuses. Figure 8.13 shows a schematic diagram of a set of resistance heaters protected by fuses. The system in figure 8.13 controls three electric heaters by energizing contactors to start the heaters. The fuses in the circuit are used as safety devices for the heaters.

Circuit breakers are used for the same purpose as fuses but allow a high starting load. However, many control circuits use fuses for protection.

NOTE: Fuses are used as overloads for H1, H2, and H3

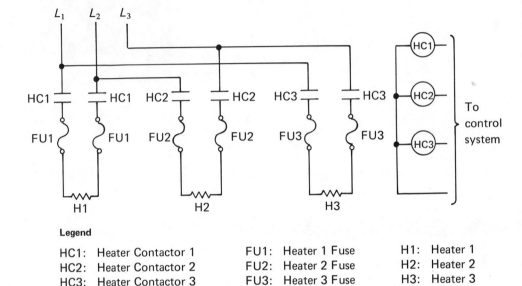

Legend

HC1:	Heater Contactor 1		FU1:	Heater 1 Fuse	H1:	Heater 1
HC2:	Heater Contactor 2		FU2:	Heater 2 Fuse	H2:	Heater 2
HC3:	Heater Contactor 3		FU3:	Heater 3 Fuse	H3:	Heater 3

FIGURE 8.13. Schematic diagram showing fuses used as overloads for protection of resistance heaters.

Line Break and Pilot Duty Overloads

Overload devices used to protect inductive loads are more effective devices than fuses but also are more complex. Inductive loads require more amperage to start than to run. The amperage of a motor at the moment power is applied is largest because the rotor of the motor is in a stationary position. Figure 8.14 shows a graph of the ampere draw of a motor from start to full speed.

Overloads can be divided into two basic groups: line break and pilot duty. The **line break overload** breaks the power to a motor. **A pilot duty overload** breaks an auxiliary set of contacts connected in the control circuit. Overloads can be manually reset or automatically reset.

Line Break Overload.   A line break overload is shown in figure 8.15. One of the most common types of line voltage overload is a metal disc mounted between two contacts. It is called a bimetal line break overload. Figure 8.16(a) shows a schematic diagram of a bimetal overload in closed position. If the current draw or temperature of the motor is sufficient to cause the disc to overheat and expand, the contact would open, as shown in figure 8.16(b). This breaks the flow of power to the load. In some overloads a

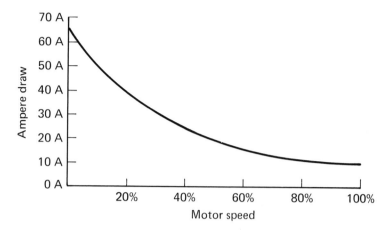

FIGURE 8.14. Ampere draw for motor speeds from locked rotor to full speed.

FIGURE 8.15. Line break overload. Photo courtesy of Texas Instruments Incorporated.

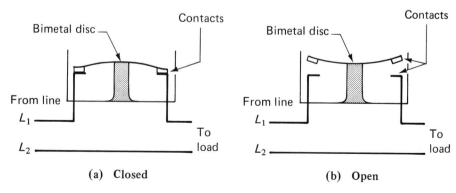

FIGURE 8.16. Schematic diagram of bimetal line break overload (a) In closed position (b) In open position.

FIGURE 8.17. Bimetal overload with a heater installed below the disc. Photo courtesy of Texas Instruments Incorporated.

heater or wire installed below the disc is sized to give off heat, as shown in figure 8.17. This gives a more accurate range of protection.

Another type of line break overload is the three-wire klixon overload. It uses the current draw of both windings to open or close the overload and break the common to both windings, as shown schematically in figure 8.18.

The most popular line break overload for use in small central residential systems is an **internal compressor overload**, shown in figure 8.19. The internal overload is a small device that is inserted into the motor windings. This overload can sense the current draw of the motor as well as the winding temperature more effectively than external overloads. Figure 8.20(a) shows a

FIGURE 8.18. Schematic diagram of a three-wire bimetal overload.

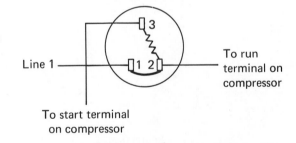

FIGURE 8.19. Internal compressor overload.

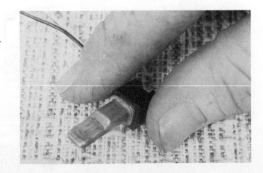

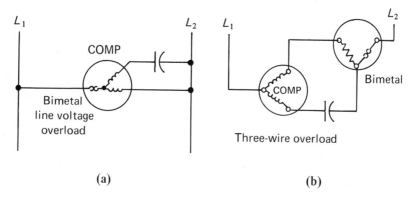

FIGURE 8.20. Schematic diagram of an internal compressor overload
(a) Bimetal overload (b) Three-wire bimetal overload.

schematic diagram of a compressor with a bimetal internal overload. Figure 8.20(b) shows a compressor with a three-wire bimetal overload. Some small three-phase hermetic motors also use an internal type of overload. Internal overloads should not be confused with internal thermostats. They are very similar in appearance. The purpose of an internal thermostat is to break the control line if the windings overheat.

Pilot Duty Overload. The pilot duty overload breaks the control circuit when an overload occurs, which would cause a contactor to be deenergized, as shown in figure 8.21. This type of overload is common on larger systems and still exists on smaller systems that are presently in the field.

FIGURE 8.21. Schematic diagram show-ing pilot duty overloads in the circuit.

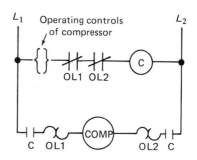

Legend

OL1:   Overload 1
OL2:   Overload 2
C: Contactor
COMP: Compressor

FIGURE 8.22. Current-type pilot duty overload. Photo courtesy of Texas Instruments Incorporated.

There are two basic pilot duty overloads being used in the industry today: the current overload and the magnetic overload. The **current overload** is shown in figure 8.22. This overload works similarly to the line break overloads except that a pilot duty set of contacts is opened rather than the line. In most cases the bimetal disc of the overload would have to be so heavy that it could not control line voltage effectively. Therefore, in larger overloads pilot duty contacts are used.

The **magnetic**, or heinemann, **overload**, as shown in figure 8.23, is another type of pilot duty overload used in the industry. The magnetic overload consists of a movable metal core in a tube filled with silicone or oil. Surrounding the metal tube is a coil of wire. When the current increases, so does the magnetic field of the coil. The overload operates by the magnetic field created by the coil. The device is designed to create a magnetic field that is strong enough to pull the core up, opening the pilot contacts on overload.

FIGURE 8.23. Magnetic overload. Photo courtesy of Heinemann Electric Co.

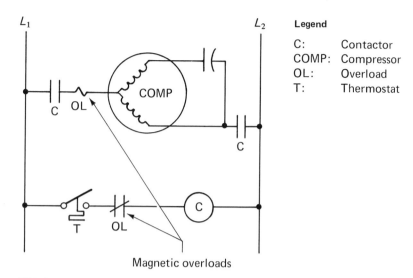

FIGURE 8.24.   *Schematic diagram showing magnetic overloads protecting a compressor.*

The magnetic overload has a time-delay feature. There is a small hole drilled in the core. Once the field begins pulling the core in, the oil or silicone must go from one end of the tube to the other through the small hole. Thus there is a short interval, due to the oil flow, between the time the motor starts up and the time the overload would break the circuit. Figure 8.24 shows a schematic diagram of magnetic overloads in a compressor circuit.

Troubleshooting

Most overloads are easy to troubleshoot because they have only one set of contacts along with a heater to heat the bimetal disc. An ohmmeter can be used to test the contacts in both pilot duty and line break overloads. In some cases there will be a break in the heater or coil, depending on the type of overload you are checking. This element of an overload can also be checked with an ohmmeter.

The internal overload of a compressor is a little more difficult to diagnose because it cannot be isolated from the system. Most internal overloads break the common of the motor. If the motor is checked and there is an open starting and running winding, chances are that the common or the internal overload is open. If the compressor housing is cool, the motor might be damaged rather than the overload. Be sure when making this test that the compressor is cool. If the compressor is warm or hot, the overload could be open and working properly.

SUMMARY

Control systems used on modern heating, cooling, and refrigeration systems use many different control components to obtain automatic control. In this chapter we described three of these components: contactors, relays, and overloads.

Contactors play an important part in the correct operation of equipment in the industry. Their purpose is to make and break a power circuit to a load, thus allowing the proper control of the equipment due to the contactor being energized or deenergized. The contacts, coil, and mechanical linkage are the main parts of the contactor. Each part must work properly for the contactor to work correctly. The coil can be faulty by being shorted or open. If it is good, a measurable resistance can be obtained on an ohmmeter. The contacts can be faulty due to wear, corrosion, or excessive friction. Whatever the diagnosis of a contactor, it should be repaired or replaced whenever it is faulty or gives erratic operation. When replacement is necessary, it is essential to select the correct contactor for the job.

The relay is used to control many loads in control systems. It is the most widely used device in the industry. The application and size of relays should be identified before any replacement is attempted. The relay has the same components as a contactor: coil, contacts, and mechanical linkage. The coil is easily diagnosed with an ohmmeter as good, open, or shorted. The contacts must be in good shape for proper operation and can be checked with an ohmmeter or voltmeter. The contacts of a relay are usually enclosed, which prevents a visual inspection. A faulty mechanical linkage in a relay usually results in sticking contacts.

Overloads play an important part in the industry because they protect expensive loads. The fuse is the simplest type of overload protection used. It is effective on noninductive loads, in protecting wires, and in protecting circuit components.

Overloads are divided into two basic types: line break and pilot duty. The line break overload breaks the line voltage to the components. It is used on small hermetic compressors and motors and is connected directly in the line voltage supply to the equipment or load. The pilot duty overload breaks a pilot duty set of contacts in the control circuit. Pilot duty overloads are arranged so that the line voltage feeds directly through them and on to the load. This line voltage will create an operating condition in the element in the overload—thermal, current, or magnetic—and open a set of pilot duty contacts. The internal overload is mounted directly in the windings of the motor and usually has no connections to the outside of the motor. The internal overload is connected in series with the common of the motor. All overloads must be properly sized to do an adequate job of protection.

QUESTIONS

1. What is the largest electric load in an air-conditioning or refrigeration system?

2. What is the major difference between a relay and a contactor?

3. The major difference between magnetic starters and contactors is that the magnetic starter _____ .

4. How does a contactor operate?

5. What are the two types of armatures used in contactors? How does each operate?

6. Name the three major sections of a contactor.

7. What is the proper procedure for checking the coil of a contactor?

8. Contacts are made of _____ .

9. What are the major reasons for replacing the contacts of a contactor?

10. True or false. A voltage reading taken across the contacts of the same pole will show the voltage drop across the contacts.

11. The easiest faults to diagnose with a contactor are problems with _____ .

12. Why is it difficult to repair contactors?

13. Name the components of a relay.

14. How does a relay operate?

15. What do the terms "normally open" and "normally closed" mean in reference to relays and contactors?

16. True or false. The contacts in a relay are easily checked by a visual inspection.

17. True or false. The size of the relay coil is smaller than the coil of a contactor.

18. What is the purpose of an overload?

19. What is a resistive load?

20. What is an inductive load?

21. Fuses are used most often to protect _____ .

22. The line break overload breaks _____ .

23. How does a line break overload operate?

24. True or false. The most popular line break overload for use in small equipment is the three-wire klixon overload.

25. The pilot duty overload breaks _____ .

26. Name the two basic types of pilot duty overloads and describe how they operate.

27. What is the advantage of an internal compressor overload?

28. What is the proper procedure for checking an internal compressor overload?

9

Thermostats, Pressure Switches, and Other Electric Control Devices

INTRODUCTION

In the preceding chapter we discussed some commonly used devices that control loads in heating, cooling, and refrigeration systems. In this chapter we will discuss devices that control some small loads and contactors and relays that control larger loads.

Thermostats are extremely important to the industry because they are opened and closed by a change in temperature. In all phases of the industry we must control temperature and in most cases we do this by use of a thermostat. Pressure switches are sometimes used to control temperature by the pressure-temperature relationship, but in most cases pressure switches are used as safety devices. Transformers are used to reduce line voltage to the low voltage that is used in many control systems.

Thermostats play an important part in almost all control systems. The thermostat is most commonly used as the primary control to control the temperature within a given area. In some cases it is also used as a safety device, as in motor protection. Thermostats that are used as primary controls can be heating thermostats or cooling thermostats or some combination of the two. Thermostats can also be used for a staging effect, that is, for operating equipment at different times depending on the demands put on the system. Staging can provide a main stage for normal operation and a secondary stage for energizing a specific part of the control system at a specific time as the load dictates.

Pressure switches can be used for different purposes. Pressure switches are often used as safety devices to cut the equipment off if the pressure is dangerously high or low. In some cases in refrigeration systems, pressure switches are used as operating controls. Whatever the purpose of the pressure switch, it always reacts to a specific pressure in a certain way.

Transformers are used to break down the incoming power voltage to a voltage that can be easily used for control circuits.

Other control devices that we will discuss in this chapter include humidistats, oil safety switches, time-delay relays, time clocks, and solenoid valves.

In this chapter we will look at each of the control devices introduced in the preceding paragraphs and will describe how they operate in the system. In the next chapter we will see how all control devices in a system can be troubleshooted.

9.1 THERMOSTATS

The temperature in any structure, regardless of its age, location, or design, can be maintained at comfortable levels with a thermostat. Thermostats are designed and built in many different forms and sizes to meet the applications required in the industry. Thermostats play an important part in the total operation of almost all systems in the industry.

Applications

The basic function of a thermostat is to respond to a temperature change by opening or closing a set of electric contacts. There are many different types of thermostats used in the industry that perform a variety of switching actions. Figure 9.1 shows several common thermostats in use today.

Thermostats are used for many different purposes. An air-conditioning or heating thermostat would basically control the temperature of a given area for human comfort. Refrigeration thermostats are designed to maintain a specific temperature within a refrigerated space, such as in a domestic refrigerator, walk-in cooler, display case, and commercial freezer. There are many types of special application thermostats used in the industry, such as outdoor thermostats and safety thermostats. Whatever use a thermostat is put to, it serves the same function: reacting to temperature with the opening and closing of a switch.

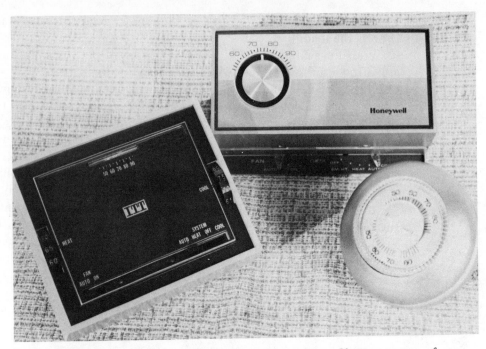

FIGURE 9.1. Some common types of thermostats. Photos courtesy of ITT General Controls and Honeywell.

A heating thermostat closes on a decrease in temperature and opens on an increase in temperature. A cooling thermostat closes on an increase in temperature and opens on a decrease in temperature. This is a very important factor to consider when ordering or installing thermostats. A heating *and* cooling thermostat is available where heating and cooling are required. The heating and cooling thermostats are usually built so that the switch reacts either for heating or for cooling, with no intermediate position; in other words, a single-pole–double throw switch. Some thermostats must isolate the heating and cooling contacts and therefore must use a separate set of contacts for heating and for cooling. Modern thermostats have a system switch that will determine whether the unit is heating or cooling.

Controlling Elements and Types of Thermostats

There are two types of **thermostat controlling elements** that are commonly used. The controlling element of a thermostat is the part that moves when a change in temperature is sensed. The bimetal thermostat shown in figure 9.2

Mercury
Bulb

Bimetal

Temperature
Setting

Thermometer

FIGURE 9.2. Internal parts of a thermo-stat. Photo courtesy of Honeywell.

is commonly used to control the temperature of air in an air-conditioning or heating application. The remote bulb thermostat is commonly used to control the temperature of any medium, whether liquid or vapor, in many applications.

Remote Bulb Thermostat.   The remote bulb thermostat is shown in figure 9.3(a). The power element, which is the bulb, and the diaphragm are interconnected with a section of small tubing. The bulb is filled with liquid and gas and then is sealed. The pressure exerted by the diaphragm on the mechanical linkage will open and close a set of contacts. A diaphragm with its remote bulb is shown in figure 9.3(b). As the bulb temperature changes, so will the pressure exerted on the diaphragm. If the temperature of the bulb increases, so will the pressure. If the temperature of the bulb decreases, so will the pressure. The increase or decrease in pressure causes the contacts to open or close, depending on the design of the thermostat.

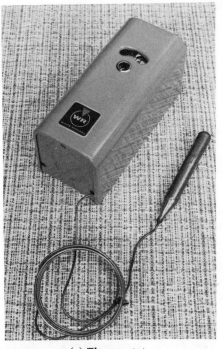

(a) Thermostat

(b) Power element and diaphragm

FIGURE 9.3. Remote bulb thermostat. White-Rodgers Div., Emerson Electric Co.

Bimetal Thermostat.   The heart of almost all other types of thermostats is a bimetal element. The element gets its name from the fact that it uses a bimetal to cause the movement that opens and closes a set of contacts. A bimetal is a combination of two pieces of metal [see figure 9.4(a)] that are a given length at a certain temperature. The two metals are welded together. Each metal has a different coefficient of expansion. If the temperature of these two metals is increased, one will become longer than the other because of the different expansion qualities. This causes the bimetal to arc, as shown in figure 9.4(b). If the bimetal is anchored at one end, leaving the other end to move freely, it will move up and down according to the temperature surrounding it. In a low-voltage thermostat the bimetal works better and gives better control if it is large.

The first type of bimetal thermostat produced is shown in figure 9.5(a). It was unsatisfactory because of the unstable pressure from the bimetal that held the contacts together. It would react (close or open) to a relatively small change in the temperature around the room thermostat.

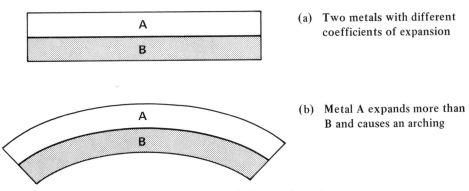

(a) Two metals with different coefficients of expansion

(b) Metal A expands more than B and causes an arching

FIGURE 9.4.   A bimetal element for a thermostat.

A thermostat must have a means of making a good connection with the contacts. This is accomplished by a **snap action** of the thermostat bimetal to the fixed contacts. The early type of thermostat was not snap-acting in the switching movement and thus caused problems because of its reaction to minor temperature change. However, if a permanent magnet is placed near the bimetal arm, as shown in figure 9.5(b), it will cause a snap action when the bimetal expands enough to move the contacts close together.

There are two methods in common use that enable thermostats to be snap-acting: a permanent magnet and a mercury bulb. A permanent magnet mounted near the fixed contacts will cause the action of the bimetal to be snap-acting. The bimetal will make good contact once it is in the magnetic field of the permanent magnet. Figure 9.6 shows this type of bimetal thermostat.

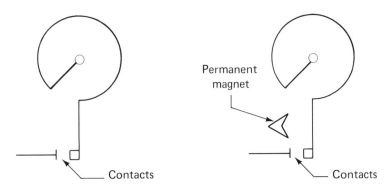

(a) Basic bimetal with no means of snap action to produce a stable set of contacts

(b) Bimetal with permanent magnet to produce snap action of the contacts

FIGURE 9.5.   Basic bimetal elements.

Contacts

85
80
75
70

FIGURE 9.6. Bimetal thermostat with snap-action contacts. Photo courtesy of Barber Coleman Co.

Bimetal

Temperature Dial

The mercury bulb thermostat also provides snap action because of the globule of mercury moving between two probes sealed inside a glass tube, as shown in figure 9.7. Figure 9.8 shows a bimetal thermostat with a mercury bulb that is used in the industry.

Line Voltage Thermostat. The line voltage thermostat is designed to operate on line voltage, for example, 110 volts or 230 volts. The line voltage thermostat, shown in figure 9.9, is used on many packaged air-conditioning units and refrigeration equipment of commercial design. This type of thermostat is used to open or close the voltage supply to a load in the system. Figure 9.10 shows the wiring diagram of a common 230-volt window air conditioner with a line voltage thermostat.

The line voltage thermostat lacks many of the good qualities that are obtained with low-voltage thermostats. For example, the line voltage thermostat functions only to open or close a set of contacts by using temperature changes. However, the line voltage thermostat is used commonly in the industry.

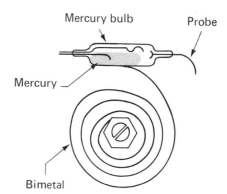

FIGURE 9.7. Bimetal with a mercury bulb used as the contacts.

Low-Voltage Thermostat. The low-voltage thermostat is used on control systems with a 24-volt supply. The low-voltage thermostat is used on all residential heating and air-conditioning systems and many commercial and industrial systems. The low-voltage thermostat can be used for heating operation, cooling operation, automatic operation of fans, manual operation of

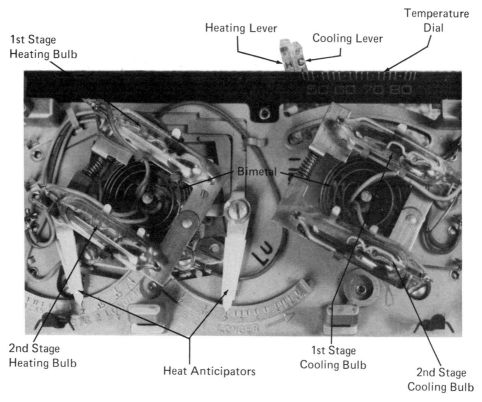

FIGURE 9.8. Bimetal thermostat with mercury bulb contacts. Photo courtesy of Honeywell.

FIGURE 9.9. Line voltage thermostat. Photo courtesy of Honeywell.

fans, and automatic changeover from heating to cooling. The difference between a line voltage and a low-voltage thermostat is in the size of the bimetal element. With the larger line voltage contacts, more pressure is needed to close the contacts and therefore larger bimetal elements are required. Figure 9.11 shows a wiring schematic of a residential heating and cooling system with a low-voltage thermostat. The low-voltage thermostat is more accurate, less expensive, and requires smaller wiring than its counterpart, the line voltage thermostat.

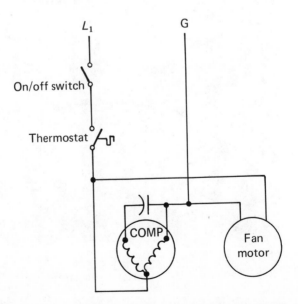

FIGURE 9.10. Wiring diagram of window unit with line voltage thermostat.

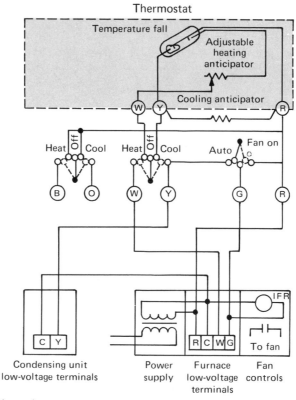

FIGURE 9.11.   Wiring diagram of a residential heating and cooling unit with a low-voltage thermostat.

Anticipators

The success of any thermostat depends on the system being correctly sized, the air flow being balanced, and the thermostat being correctly placed. It is almost impossible for thermostats to maintain the exact temperature of any given area without spending an unreasonable amount for controls. Thus **anticipators** are used to give a more evenly controlled temperature range. There are two types of anticipators: heating and cooling. We will discuss each type in turn.

Heating Anticipator.   A thermostat that has no means of heat anticipation

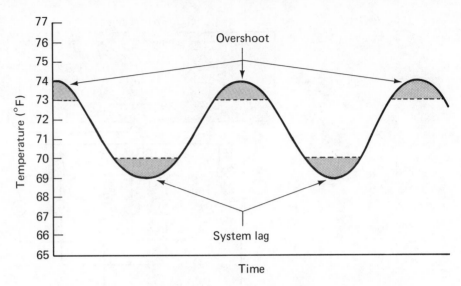

FIGURE 9.12.   *Effect of overshoot and system lag on room temperature.*

will allow a wide swing from the desired temperature, especially on forced-warm-air systems. If the thermostat is set on 75°F, the furnace will come on for heating. But there is a delay before any warm air is carried to the conditioned space because a warm-air heating system must heat the furnace first. This delay will allow the air temperature to drop to 74°F or less before the blower begins operation and heat is carried to the conditioned space.

The temperature difference between the closing of the thermostat and the time when warm air begins to reach the thermostat is called **system lag**. As long as the furnace remains on, the temperature will continue to rise. If the thermostat differential is 2°F, the thermostat will open at 77°F and stop the burner. However, the furnace is still warm and the blower must be operated until the furnace cools, which will carry the temperature to 78°F or higher.

The temperature difference between the closing of the thermostat and the time when the warm air is no longer being delivered to the room is called **overshoot**. The overshoot and system lag can produce an additional temperature swing of as much as 5°F. Figure 9.12 shows the effect of overshoot and system lag.

The wide difference in temperature can be controlled by using a heating anticipator in the thermostat. The heating anticipator is nothing more than a small source of heat close to the bimetal element. This allows the bimetal to be a little warmer than the surrounding air. The heating anticipator is placed in series with the contacts of the thermostat. When the contacts close and

energize the burner, the current must also flow through the heating anticipator, which causes it to heat the bimetal to a certain temperature.

A heating anticipator anticipates the point at which the thermostat
should open and provides a narrow thermostat differential. Suppose the
thermostat is set at 75°F and the furnace burner is off, with the temperature
dropping slowly. When the temperature falls below 75°F, the burner starts
and at the same time the heat anticipator begins to heat the bimetal. The air
temperature will continue to drop until the blower delivers warm air to the
conditioned space. The room temperature begins to rise, but the bimetal
temperature is slightly higher because of the heat from the heating anticipator. Thus the thermostat anticipates the overshoot and cuts off the burner.
Due to the shorter length of time that the burner operates, the furnace will
cool quicker, stopping the fan and preventing the overshoot. A heating anticipator does not eliminate system lag and overshoot. But these factors become
negligible with a heating anticipator, producing a temperature swing of only
2°F to 3°F. The thermostat circuit in figure 9.11 shows a heating anticipator
connected in series with the contacts of the thermostat.

There are two types of heating anticipators used: fixed and adjustable.
The fixed heating anticipator is nonadjustable and is not very versatile. The
adjustable heating anticipator, shown in figure 9.13, can be matched to
almost any control system. The adjustable heating anticipator should be set
on the current draw of the primary control or the gas valve. Matching the
adjustable heating anticipator to the current rating of the burner control
assures the best possible heat anticipation.

Cooling Anticipator.   The cooling anticipator operates somewhat differently
from the heating anticipator. This type of anticipator is also known as an
off-cycle anticipator. The cooling anticipator is shown connected to a
thermostat in figure 9.14. The cooling anticipator adds heat to the bimetal
on the "off" cycle of the equipment because of its parallel connection in the
circuit. When the thermostat contacts close, the current takes the flow of
least resistance, which is through the contacts rather than the cooling antici-

*FIGURE 9.13. Adjustable heating antic-
ipator. Photo courtesy of Honeywell.*

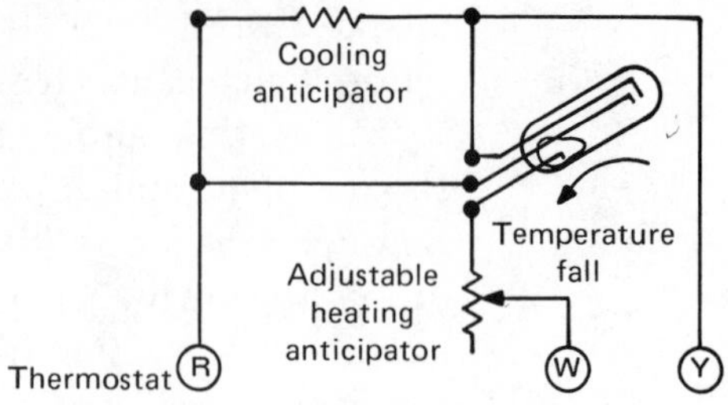

FIGURE 9.14. Schematic diagram of cooling anticipator connected to a thermostat.

pator. On the "off" cycle the current passes through the anticipator and the contactor coil. The anticipator drops the voltage to a point that will not energize the contactor, because of the series connection. The current flow through the anticipator heats the bimetal and causes the thermostat to anticipate the temperature rise. The cooling anticipator is not as important as the heating anticipator because cooling equipment operation is almost instant, while there is a delay in the heat delivery of a forced-warm-air system.

The thermostat is merely a switching device that routes the voltage to the correct control for the operation prescribed by the thermostat setting. Figure 9.15 shows the schematic of a low-voltage thermostat with its routing. With the system selector in the cooling position, voltage flows through the system switch to the cooling thermostat, which operates the cooling equipment and the fan motor to maintain the temperature setting. If the system switch is in the heating position, the current flows through the switch to the heating thermostat, which operates the heating equipment. Many thermostats incorporate a fan switch that allows the fan to be operated manually or with the cooling equipment, since the fan on heating operates from a fan switch.

Thermostat Installation

Installation of a thermostat is fairly simple because all thermostats are marked with identifiable letters, although the letters used are not consistent from manufacturer to manufacturer. Figure 9.16 shows a simple heating and cooling thermostat subbase. The terminals are shown with their letter designations (the thermostat attaches to the subbase). The letter designations to system functions are given in the chart in figure 9.16. The letter designation should be followed for proper installation.

When checking thermostat operation, check the wiring connections and make sure that the selector switch and the temperature setting are properly selected.

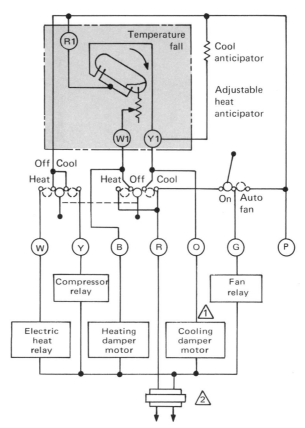

 With no O terminal load, thermostat current during heating cycle varies depending on whether fan switch is in the On or Auto position. Heater should be set for combined current level of heat relay and fan relay coils. With O terminal load, set thermostat heat anticipator to its maximum setting as cooling anticipator in series with O terminal load provides heat anticipation in the heating cycle. (Limit the thermostat heating load current to 0.8 amps to assure good performance.)

△2 Power supply provide overload protection and disconnect means as required.

FIGURE 9.15. Schematic diagram of the routing of a low-voltage thermostat. Diagram courtesy of Honeywell.

9.2 STAGING THERMOSTATS

A **staging thermostat** is designed to operate equipment at different times with respect to the equipment's needs (see figure 9.8). The staging thermostat has more than one contact and opens and closes at different times with regard to the condition of the area being controlled.

199

Letter Designation

Voltage to Transformer	R–V
Heating	W–H
Cooling	Y–C
Fan	G–F
Heating Damper (seldom used)	B
Cooling Damper (seldom used)	O

FIGURE 9.16.   Thermostat subbase with terminals and letter designations. Photo courtesy of Honeywell.

The heating, cooling, and refrigeration industry has advanced in almost all phases, but in recent years there has been a demand for better control and more efficient operation with the modern systems. Staging thermostats have been designed to meet this need. The staging thermostat is becoming increasingly popular in the industry because of its versatility in system control.

Staging System

Many heating and cooling systems are operated in stages because the load in some structures fluctuates a great deal. A heating or cooling system that has been designed to operate on two different capacity levels is a **staged system**. A staged system is designed to operate at half of its capacity or more until the operating section of the equipment can no longer handle the heating or cooling needs of the structure. Then additional stages are called on as they are needed.

Staging systems offer many advantages because of their more efficient operation. For example, on a mild day the cooling load of a building is low and the full capacity of the equipment is not needed. Hence the equipment will operate at half of its capacity. If the day is extremely hot and the full capacity of the system is needed, then it can be utilized by use of a staging thermostat. Staging can be used to many advantages whether it is in heating or cooling.

Operation and Types

A staging thermostat is designed to be used on a system that has two stages of heating, cooling, or both. The staging thermostat operates on the differential in temperature between the stages. For example, a two-stage-cooling thermostat could close one set of contacts at 75°F and the other set at 76.5°F to 78°F. A two-stage-heating thermostat could close one set of contacts at 75°F and the other set at 73.5°F to 72°F. This allows systems to operate on partial capacity until the need arises for full capacity.

Staging thermostats can be obtained in a variety of stage configurations. Common staging thermostats used in the industry are the one-stage-heating thermostat with two-stage cooling, the two-stage-heating thermostat with one-stage cooling, and the two-stage-cooling thermostat with two-stage heating. Figure 9.17 shows a two-stage-heating and two-stage-cooling thermostat. Note that four mercury tubes are used, with two mounted on the cooling

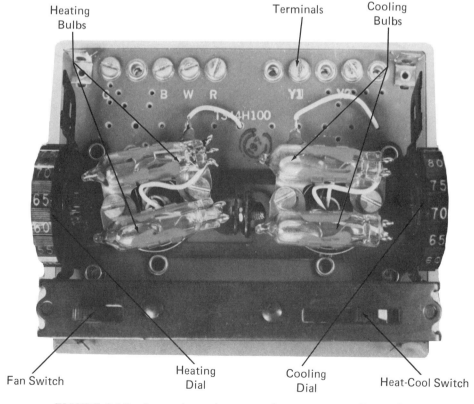

FIGURE 9.17. Low-voltage thermostat for two-stage cooling and two-stage heating. Photo courtesy of ITT General Controls.

201

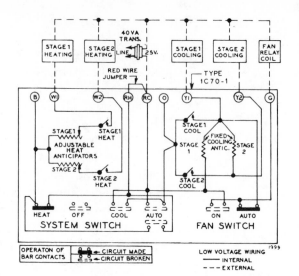

FIGURE 9.18. Schematic diagram for a two-stage-cooling and two-stage-heating thermostat. White-Rodgers Div., Emerson Electric Co.

control level and two mounted on the heating control level. Figure 9.18 shows the schematic diagram for the two-stage-heating and two-stage-cooling thermostat.

The letter designations are the same for staging thermostats as they are for regular thermostats except that heating or cooling stages would be lettered with a 1 or 2 after the main letter. The number 1 represents the first stage; the number 2 represents the second stage. When installing staging thermostats, pay careful attention to the letter designations and control hookup to avoid an improper operating system.

Heat Pump

A **heat pump** is a refrigeration system that heats or cools by reversing the refrigerant cycle for the heating operation and then operates conventionally for cooling. Most heat pumps use a staging thermostat to operate a set of supplementary heaters. The first-stage thermostat operates the compressor. The second stage operates the supplementary heat when the first stage cannot handle the load. Figure 9.19 is a wiring diagram of a heat pump showing the two stages of heating and what they control.

9.3 PRESSURE SWITCHES

A pressure switch is a device that opens or closes a set of contacts when a certain pressure is applied to the diaphragm of the switch. A high-pressure switch is connected to the discharge side of the system to sense discharge

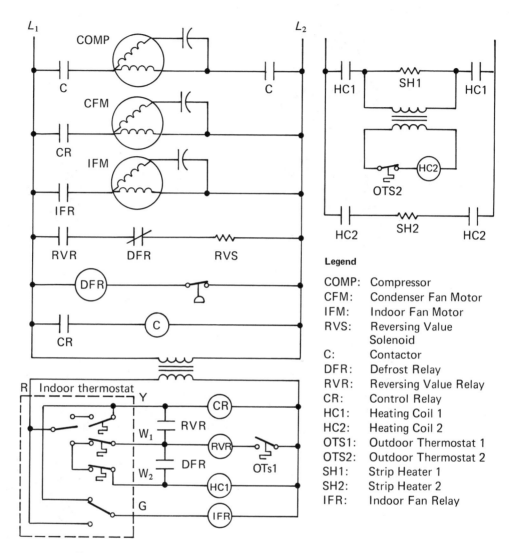

FIGURE 9.19. Schematic diagram of a heat pump using a two-stage-heating and one-stage-cooling thermostat to control the unit.

pressure. A low-pressure switch is connected to the suction side of the system to sense suction pressure. Pressure switches can be used as safety devices, as main operating controls, or to operate other parts of the system. Figure 9.20(a) shows a common pressure switch used in the industry.

Two types of pressure switches are used in the industry today. A non-adjustable pressure switch is used by many manufacturers to prevent the pressure setting from changing. An adjustable pressure switch can be adjusted to meet any specific need that might arise. Adjustable pressure

(a) With enclosure   (b) Without enclosure

FIGURE 9.20.   Pressure switches (a) with enclosure (b) without enclosure. Photo (a) courtesy of Penn Division, Johnson Controls, Inc.

switches are usually obtained for field replacement of a pressure switch. Pressure switches can also be classified as high pressure or low pressure. These switches open or close on a rise or fall of pressure. Pressure switches are manufactured with a case enclosing the switch, as shown in figure 9.20(a), or for mounting in a control panel without the case, as shown in figure 9.20(b). A dual-pressure switch contains both a high-pressure and a low-

FIGURE   9.21. Dual-pressure   switch (combination of high and low pressure). Photo courtesy of Penn Division, Johnson Controls, Inc.

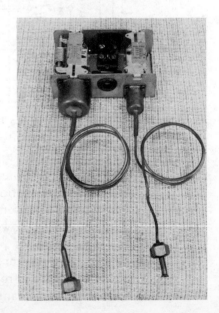

FIGURE 9.22. Table of approximate setting points of a high-pressure switch used as a safety control.

Refrigerant	Condenser Type	High Pressure Setting	
		Cut-out	Cut-in
12	Air Cooled	225	145
	Water Cooled	170	90
500	Air Cooled	280	200
	Water Cooled	210	130
22	Air Cooled	380	300
	Water Cooled	280	200

pressure switch, as shown in figure 9.21. Pressure switches can be obtained for almost any purpose that would arise in the industry.

High-Pressure Switch

A high-pressure switch is usually used as a safety device to protect the compressor and system from excessively high discharge pressure. A high-pressure switch used as a safety control must open on a rise in pressure to shut the equipment down.

The high-pressure setting of the switch must correspond to the type of refrigerant in the system. The setting of a high-pressure switch used with Freon 12 would be different from the setting of a high-pressure switch used with Freon 22. Figure 9.22 lists the common setting points of a high-pressure switch for the common refrigerants used in the industry today. High-pressure switches may also close on a rise in pressure to operate a device that would control the discharge pressure.

Low-Pressure Switches

Low-pressure switches are used as safety devices, as operating controls, and as devices to operate any component by the suction pressure of the system. All low-pressure switches are connected to the suction side of a refrigeration system. Low-suction pressure can cause damage to the compressor. Therefore, low-pressure switches are used many times as a safety device to prevent damage to the system when the suction pressure drops below a predetermined point.

Figure 9.23 lists the common setting points of a low-pressure switch for different types of Freon.

FIGURE 9.23. Table of approximate setting points of a low-pressure switch used as a safety control.

Refrigerant	Low Pressure Setting	
	Cut out	Cut in
12	15	35
500	22	46
22	38	68

If the low-pressure switch opens, it should break the control circuit that operates the compressor. Low-pressure switches can also be used as an operating control to operate the system by a pressure setting that corresponds to a temperature setting. Therefore, by properly setting a low-pressure switch, the temperature can also be controlled. Figure 9.24 lists the proper pressure control settings for specific applications.

Notation and Terms

There are many terms that should be understood to successfully maintain and set pressure switches. The **differential** of a pressure switch is the difference between the cut-in and cut-out pressure of the switch. The **cut-in** is the pressure of the system when the pressure switch closes. The **cut-out** is the pressure of the system when the pressure switch opens. The **range** of a con-

	REFRIGERANT					
	12		22		502	
Application	In	Out	In	Out	In	Out
Walk-in Cooler	32	15	60	35	72	40
Dairy Case–Open	35	12	64	24	75	30
Reach-in Display Case–Open	36	14	66	33	78	40
Meat Display Case–Closed	35	15	65	34	78	45
Meat Display Case–Open	30	12	55	30	65	35
Vegetable Display–Open	40	15	80	36	90	45
Beverage Cooler–Wet Type	30	20	58	38	68	58
Beverage Cooler–Dry Type	35	15	65	35	76	43
Florist Box	45	28	80	55	85	63
Frozen Food–Open	6	6	16	4	25	10
Frozen Food–Closed	10	2	22	12	30	15
Walk-in Freezer (–10°F)	12	1	32	10	25	15

Note: Italicized numbers represent vacuum: Bold numbers represent PSIG
Caution: The figures are approximate and should be used only as a guide. Special applications should not be considered.

FIGURE 9.24.   Table of approximate pressure control settings of a low-pressure switch used as an operating control.

trol is the operating range of the system—for example, the overall pressure—over which the switch can operate. These terms are used frequently when setting pressure to obtain the correct operation of the system. Many pressure controls can be set by carefully reading the pressure dial of a pressure switch. Care should be taken when setting or replacing switches.

Troubleshooting

There are many problems commonly encountered when working with pressure switches. The contacts of pressure switches frequently cause problems because of pitting, wear, mechanical linkage faults, and sticking. The contacts should be checked to see if they are closed or open in relation to the setting of the control.

If the contacts show continuity, the resistance should also be checked. If the resistance is above allowable limits, or if the contacts prove to be faulty, the pressure switch should be replaced. If any resistance is read through a set of contacts, voltage is being lost. A reading of 2 or 3 ohms of resistance in a set of switches indicates that the switch should be replaced or at least that the voltage should be checked across the contacts.

The pressure connections of a pressure switch can leak and cause a faulty control or no control at all. To check for a leak use an acceptable method for locating Freon leak. Never replace a pressure switch before its correct position and purpose have been determined.

9.4 TRANSFORMERS

The transformer in a heating or cooling system provides the low-voltage power source for the control circuit. It transforms line voltage to the suitable low voltage that is needed for the control system. Most residences and small commercial installations incorporate a 24-volt control system. The transformer for a residential unit, shown in figure 9.25, is used to transform the voltage from line voltage to 24 volts. Some commercial and industrial high-voltage equipment will use transformers like the one shown in figure 9.26 to drop voltage from the supply to a 230-volt or a 110-volt control system.

Operation

Transformers are stationary inductive devices that transfer electric energy from one circuit to another by induction. The transformer has two windings,

FIGURE 9.25. Transformer used in residential and small commercial systems. Photo courtesy of Penn Division, Johnson Controls, Inc.

primary and secondary. An alternating voltage is applied to the primary winding of a transformer and induces a current in the secondary winding.

There are many different types of transformers used in the industry. A step-down transformer induces a secondary voltage at a lower rate than the primary. This type of transformer is used for the power supply of a low-voltage system. A step-up transformer induces a secondary voltage at a rate higher than the primary. This type of transformer is used to boost the voltage.

FIGURE 9.26. Transformer used in large commercial and industrial units.

Sizing Transformers

Transformers are like many other electric components. They are not 100% efficient. In other words, there is a loss between the primary and secondary windings. This loss must be considered when sizing transformers for a certain job. Transformers are rated in voltamperes (VA). System equipment must be considered in transformer sizing along with the transformer rating.

The selection of a transformer vitally affects the performance and life of electric components in heating, cooling, and refrigeration equipment. A transformer too small for the control circuit will result in a lower-than-normal low voltage to the control circuit. This will result in improper operation of contactors or motor starters due to chattering or sticking contacts, burned holding coils, or the failure of contacts to close properly. All these conditions can cause system failure and possible damage to the compressor.

Even when transformers are sized correctly, care should be taken to avoid an excessive voltage drop in the low-voltage control circuit. When using a 24-volt control system with a remote thermostat, size the thermostat wire to carry sufficient current between the transformer and the thermostat.

The capacity of a transformer is described by its electric rating, primary voltage, frequency, secondary voltage, and the load rating in voltamperes. The maximum load of a transformer used in an air-conditioning control circuit is 100 VA. The maximum voltage is 30 V. Fuses can be used on the secondary side of the transformer for protection.

Transformers should be selected so they will operate all 24-volt loads without overheating. Transformers used only on furnaces will generally be rated 20 VA or less because of the light low-voltage loads. Air-conditioning systems usually require a transformer rated at 40 VA or larger, depending on the low-voltage loads. When replacing transformers in the field, use the following guidelines:

1. For replacement select a transformer the same size or larger than the one being replaced.
2. For new applications follow the manufacturer's recommendation.

Troubleshooting

Transformers can be easily diagnosed by two methods. An ohmmeter can be used to check the continuity of the windings of a transformer for an open, shorted, or good condition. However, it is difficult to diagnose a transformer with a spot burnout unless the second method of diagnosis is followed. A voltmeter can be used to check the secondary voltage of a transformer with the correct line voltage applied to the primary. When checking with a voltmeter, there should be no load applied to the transformer. In some cases when a load is applied, the transformer secondary drops below an acceptable level, but this condition is rare.

When replacing a faulty transformer, the low-voltage control should be checked, because a short in the circuit will cause the transformer to burn out again. This is a common occurrence when a contactor coil or relay coil shorts out.

9.5 MISCELLANEOUS ELECTRIC COMPONENTS

Humidistats

In certain air-conditioning applications it is essential to control the humidity. **Humidistats** are used to control the humidity of a structure. The humidistat uses a moisture-sensitive element to control a mechanical linkage that will open or close an electric switch according to the humidity. Figure 9.27 shows a humidistat enclosed and with the cover removed.

In certain fiber and cotton industries the humidity must be accurately controlled by a humidistat, which controls the operation of air washers to prevent damage to the finished product. Humidity control is also advisable in the winter in many structures to produce a more comfortable area by adding moisture to the air.

FIGURE 9.27. Humidistat. Photos courtesy of Honeywell.

(a) Enclosed   (b) Without cover

Oil Safety Switch

Oil safety controls are essential on commercial and industrial equipment to provide adequate protection for the compressor in case the oil pressure drops. Large compressors use a pressure type of lubrication system that must be maintained at a given pressure to ensure proper lubrication of the compressor.

To correctly read the oil pressure of a compressor, you must subtract the suction pressure from the oil pressure to get the net oil pressure, because the suction pressure exerts pressure on the oil in the crankcase. Therefore, the pressure connections of an oil safety switch should be connected to the oil pressure port on the compressor and to the suction or crankcase pressure. These connections of pressure to the oil safety switch will transfer the net oil pressure to the control by some type of mechanical linkage to open and close a set of contacts, depending on the oil pressure setting. Figure 9.28 shows an oil safety switch and its schematic diagram.

The oil safety switch is designed to allow a certain time delay so that the oil pressure will build up in the compressor after the start-up of the compressor. When the compressor is energized, the time-delay switch in the oil safety control is also energized. If the oil pressure does not reach a certain level within the time-delay period, the control circuit will be deenergized. However, if the oil pressure reaches the desired level, the time-delay switch is removed from the circuit and the compressor continues to operate. The time-delay switch is nothing more than a heater that opens a bimetal switch after the specified period of time delay.

Oil safety switches are rated for pilot duty and in most cases must be

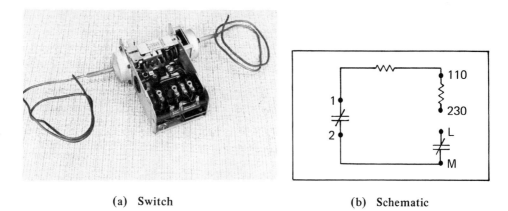

(a) Switch   (b) Schematic

FIGURE 9.28. Oil safety switch. Photo courtesy of Penn Division, Johnson Controls, Inc.

FIGURE 9.29. Time-delay relay. Photos courtesy of Essex Group, Controls Div. and White-Rodgers Div., Emerson Electric Co.

manually reset. Oil safety controls are essential on large expensive compressors to prevent undue damage because of inadequate lubrication.

Time-Delay Relay

Time-delay relays are used in the industry to delay the starting of some load for a designated period of time. Time-delay relays usually use some type of heating element that closes a bimetal element. The time-delay function is brought about by the period of time that it takes the heating element to open or close the bimetal. One type of time-delay relay is shown in figure 9.29. The relay contacts are usually rated as pilot duty. The coil or heater of the relay can be rated at 24 volts, 110 volts, or 230 volts.

Time-delay relays can be used to prevent two heavy loads of a system or two units connected in parallel from starting at the same time by putting the contacts of the time-delay relay in series with the load or controlling element of the unit. Some commercial and industrial units use a part winding motor, which uses a time-delay relay to energize the windings at different times. The delay period of a time-delay relay can vary from 15 seconds to 30 seconds or longer.

Time Clock

Time clocks are devices that open and close a set of contacts at a selected time by a mechanical linkage between the contacts and a clock. Time clocks can operate on a one-day basis or a seven-day basis. The one-day time clock, shown in figure 9.30, operates on a 24-hour period of time without regard to the day of the week. A seven-day time clock operates on an hourly or a daily basis during a week.

Time clocks are set by attaching clips to the clock wheel or dial, as shown in figure 9.30. There are two clips that attach to the wheel or dial. The A clips attach to the dial and, on movement of the dial, they close the contacts at the point in time where they are attached to the dial. The B clips open the contacts at the point in time where they are set on the dial, on movement of the dial.

Time clocks are used to operate equipment or sections of equipment on a time basis. One of the most common uses of a time clock is to stop and start the defrosting cycle of a refrigeration system. Time clocks are also used to cut systems off when a structure is to be unoccupied and to cut systems in before building occupants arrive. Buildings such as churches, offices, or manufacturing plants that close and open at specific times often use time clocks. Time clocks can be used effectively to save the operating cost of equipment when it is not needed.

Solenoid Valves

Solenoid valves are valves that open and close due to a magnetic or solenoid coil being energized, which pulls a steel core into the magnetic field of the solenoid. A common solenoid is shown in figure 9.31. Solenoid valves stop or start the flow of a fluid such as water, air, or refrigerant. Solenoids may be normally open or normally closed and care should be taken to identify the normal position of the valve.

In this section only a few miscellaneous electric controls have been covered, but there are many others used in the industry for special purposes.

FIGURE 9.30.   Clips attached to the dial of a one-day time clock. Photo courtesy of Paragon Electric Co., Sub. of AMF Inc.

FIGURE 9.31. A common solenoid valve. Photo courtesy of Honeywell.

Most controls not covered can be understood by studying the device and reading the manufacturer's instructions on the device. Service personnel should understand all basic electric components and their operation in order to correctly diagnose problems in components. There are some devices that are difficult and complex but in most cases manufacturers will assist mechanics in becoming familiar with special devices by their literature and through service schools.

SUMMARY

In this chapter we discussed several different types of control devices, among them thermostats, pressure switches, and transformers.

Thermostats are designed to open and close a set of electric contacts. They are used to control temperature in a structure or the temperature of a medium such as a liquid or gas. There are two types of thermostats used in the industry today. The remote bulb thermostat is used for control of any medium and is usually used in commercial and industrial applications. The bimetal thermostat is used in all residences and in many cases when air temperature is controlled in commercial and industrial applications. Thermostats can also be of the line voltage or low-voltage type. The line voltage thermostat is used to open or close the voltage supply to a load in a system. The low-voltage thermostat is often used as a switching device to control low-voltage loads.

The heating thermostat that does not have a heating anticipator will allow the temperature of the controlled area to have a large swing in differential. Heating anticipators, when incorporated with a heating thermostat, will have a small temperature differential. Heating anticipators are connected in series with the contacts of the thermostat and set to match the burner control. Cooling anticipators are connected in parallel and operate when the equipment is off. They anticipate the need of the equipment to operate sooner because of system lag. Thermostats play an important part in the industry because of their function of temperature control.

Staging thermostats are used to provide a means of controlling two stages of heating or cooling with a set differential. The staging of equipment gives better comfort and more efficient operation of equipment. Staging is used almost exclusively in commercial and industrial applications.

Pressure switches are widely used devices in the industry. They are often used to control a load in a system. Pressure switches are designed to open or close on a rise or fall in pressure, depending on their application.

Low-pressure switches are often used in the industry to protect refrigeration components from low-suction pressure. Low-pressure switches are used to operate systems or components and to act as safety devices to protect the refrigeration components of the system.

High-pressure switches are used as safety devices to protect the refrigeration system from excess pressure or to operate a component for head pressure control. High head pressure can damage a compressor, metering device, and high-side system piping. High-pressure switches can also be used to control loads that must be started from the discharge pressure. Careful attention should be paid to the type of system and refrigerant when checking the pressure setting of a control.

Transformers decrease or increase the applied voltage to the desired voltage by use of induction. Transformer ratings include primary voltage and frequency, secondary voltage, and voltamperes. Capacity is rated in voltamperes, that is, the voltage times the current of the control circuit loads. The more low-voltage loads in a control circuit, the larger the transformer that must be used.

QUESTIONS

1. What is the purpose of a thermostat?

2. The two types of thermostats are _____ and _____.

3. What is the difference in the operation of a heating thermostat and a cooling thermostat?

4. The controlling element of a thermostat is the part _____.

5. Explain the operation of the remote bulb thermostat.

6. True or false. A bimetal is a combination of two pieces of metal welded together.

7. Why should the contacts of a thermostat be snap-acting?

8. Name the two methods in common use that enable thermostats to be snap-acting.

9. True or false. The line voltage thermostat reduces the line voltage being supplied to the load.

10. The low-voltage thermostat is used on _____.

11. The difference between a line voltage and a low-voltage thermostat is in _____.

12. What are the letter designations R, Y, Y1, Y2, W1, W2, and G used for in thermostats?

13. What is the purpose of a heating anticipator?

14. What is the proper procedure for checking thermostats?

15. Describe what is meant by system lag and overshoot.

16. What is the purpose of staging heating and cooling equipment?

17. Why are two-stage-heating thermostats used on heat pumps?

18. What is the purpose of a low-pressure switch?

19. What is the purpose of a high-pressure switch?

20. How can a low-pressure switch be used as an operating control on a refrigeration system?

21. Define the following terms: cut-in, cut-out, range, and differential.

22. What is the purpose of a transformer and how does it work?

23. True or false. A step-down transformer is used to boost voltage.

24. What is the correct procedure for checking transformers?

25. The capacity of a transformer is rated in _____.

26. What precautions should be taken when replacing transformers?

27. What is the purpose of a humidistat?

28. The oil safety control operates on what two pressures in the refrigeration systems?

29. Time-delay relays are used in heating, cooling, and refrigeration systems to _____.

30. Give several reasons for using a time clock in the heating and cooling industry.

31. Why is it important for service personnel to understand the operation of any electric component in the system?

10

Troubleshooting Electric Control Devices

INTRODUCTION

Most troubleshooting in a system involves a specific problem that the consumer is encountering. Most problems in a system stem from one source—an electric component that is not functioning properly or is faulty. It is the responsibility of the service mechanic to locate the component that is not functioning correctly and to replace it or repair it. It is sometimes difficult to locate the exact trouble in the entire system, but the task should be relatively simple once the problem has been narrowed down to a single or several components. In this chapter we will discuss troubleshooting for most of the basic electric control components.

The first step in troubleshooting any component is to understand its operation and function. If the operation of an electric component is not understood, it is impossible to effectively check the component. Electric meters will usually be needed in the diagnosis of the component. Thus it is essential for service mechanics to understand the use of electric meters. Service mechanics and other personnel must also understand the proper procedures for checking electric components and be able to correctly diagnose the condition of the component. In the following sections we will discuss some guidelines to use in checking electric components and in diagnosing problems of components in modern heating, air-conditioning, and refrigeration systems.

10.1 CONTACTORS AND RELAYS

Contactors and relays are used on almost all heating, cooling, and refrigeration equipment for the operation of loads in the system. Contactors and relays are similar in their operation, because both contain sets of contacts and a coil that is used to open or close the contacts. The contactor is larger and capable of carrying more amperage than the relay.

The same procedure can be used to check both contactors and relays. There are three areas of problems encountered with contactors and relays: the contacts, the coil, and the mechanical linkage. Any one of the three areas can cause a contactor or relay to malfunction.

Contacts

The contacts of a relay or contactor must make good direct contact when energized for the device to function properly. One of the problems often encountered with contactors and relays is the contacts' inability to make good contact. The contacts can be burned, pitted, or stuck together. A set of contacts that are burned or pitted can cause a voltage drop across the contacts.

There are several methods of checking a set of contacts to determine if they are burned or pitted enough to warrant changing the device. The easiest method is to make a visual inspection. Figure 10.1 shows a contactor with a severely damaged set of contacts. Most contactors have movable covers, which allow easy visual inspection. Most relays are sealed and visual inspection is impossible.

A resistance check can also determine the condition of a set of contacts. The device must be energized to check normally open contacts. Normally closed contacts must be checked with the device deenergized. If the resistance is greater than 2 or 3 ohms, the contacts should be considered faulty.

FIGURE 10.1. Contactor with damaged contacts.

A voltage check can also determine the condition of a set of contacts. When a voltage check is made, the contactor or relay should be energized. To make a voltage check on a set of contacts, take a voltage reading from one side of the contacts to the other. The reading will show how much voltage is being lost. The load must be energized when a voltage test is being performed. If the voltage loss across the contacts exceeds 5% of the line voltage, the contacts are faulty and the contactor or relay should be replaced.

The contact alignment can also cause a contactor or relay to malfunction. Contacts should close directly in line with each other and seat directly in line with good firm contact. The major cause of contact misalignment is a faulty mechanical linkage. If the contacts are out of alignment, the contactor or relay must be rebuilt and any faulty components replaced.

Coil

The coil of a relay or contactor is used to close the contacts by creating a magnetic field that will pull the plunger into the magnetic field. If the coil of a relay or contactor is faulty, the device will not close the contacts. A coil of a contactor or relay should be checked for opens, shorts, or a measurable resistance. If a coil is shorted, the resistance will be 0 ohms and the coil should be replaced. An open coil will give a resistance reading of infinity and this coil should also be replaced. A measurable resistance indicates that the coil is good. Almost any measurable resistance indicates a good coil because of the variance in coil voltages. A shorted contactor coil will cause a transformer to burn out, and the service mechanic should take caution not to allow this to happen.

Mechanical Linkage

The mechanical linkage of a contactor or relay can cause malfunctions in many different forms, such as sticking contacts, contacts that will not close due to excess friction, contacts that do not make good direct contact, and misalignment of contacts. The best method for detecting a faulty mechanical linkage is by visual inspection. On contactors and some relays this can be done by removing the device and merely looking it over. However, most relays are sealed and inspection is impossible. A sealed relay must be checked by determining if the contacts open and close when the relay coil is energized or deenergized.

A mechanical linkage problem can cause a contactor or relay to stick open or closed or cause misalignment of the contacts. If a contactor or serviceable relay has a mechanical linkage problem, it should be replaced (unless it can be easily repaired).

10.2 OVERLOADS

Most major loads used in heating, cooling, and refrigeration equipment have some type of overload protection. Overloads are often overlooked as being a problem in the system but they may be faulty. A faulty overload can cause the equipment to run without protection or not operate at all. The high cost of the major loads in a system makes it necessary to protect all major loads.

Fuse

The fuse is the easiest type of overload to check because of its simplicity. A fuse can easily be checked with an ohmmeter in most cases. If a 0-ohm resistance is shown, the fuse is good. No continuity indicates a bad fuse.

A fuse on rare occasions will not completely blow or break but will partially burn out. In this case the fuse will show 0 ohms but will not allow enough current through it to operate the load. A voltage check across each fuse while power is applied to the load will show a partially burned-out fuse. The voltage check is done by placing the leads of a voltmeter across each end of the fuse. If line voltage is read on the meter, the fuse is bad and should be replaced. If no voltage is read, the fuse is good.

Circuit Breaker

The circuit breaker is another type of overload device that is used by some equipment manufacturers and in many electric panels that are in common use in the industry today. The circuit breaker is a device that will trip or open on an overload and must be manually reset. The circuit breaker is checked by taking a voltage reading on the load side of the circuit breaker, as shown in figure 10.2. If line voltage is read on the load side of a circuit breaker, it is probably good.

Circuit breakers can also cause trouble by tripping at a lower amperage than its rating. An amperage check of a circuit breaker is done by using an ammeter. If the circuit breaker trips at a lower amperage than its rating, it should be replaced. A circuit breaker can also cause nuisance trippings if it is unable to handle its rated amperage.

FIGURE 10.2. Checking a circuit breaker with a voltmeter. Photo courtesy of General Electric Co., Circuit Protective Devices Products Dept.

Line Voltage Overload

A line voltage overload installed on a load device is the easiest type of overload to check. It is used on small hermetic compressors and motors and is connected directly to the line voltage supply. The line voltage overload can be open, permanently closed, or open on a lower-than-rated ampere draw. A line voltage overload has only two or three terminals to check. An ohmmeter across the terminals will indicate whether the overload is open or closed.

Care should be taken not to condemn an overload when, in fact, it is open because of a malfunction of the load it is controlling. If a line voltage overload is weak, an amperage check should be made to see what amperage

221

is causing the overload to open. If the dropout amperage is lower than the overload rating, the device should be replaced.

Pilot Duty Overload

A pilot duty overload has a set of contacts that will open if an overload occurs in the line voltage side of the overload. These overloads are arranged so that the line voltage feeds directly through them and then on to the load. The line voltage section of a pilot duty overload can be controlled by heat, amperage, or magnetism—all three are in common use in the industry today. This type of overload is harder to check than the overloads discussed previously because of its complexity. The pilot duty contacts and the controlling line voltage element must be checked.

The pilot duty contacts on a pilot duty overload are easy to check by using an ohmmeter. The pilot contacts are usually easy to distinguish from the line voltage components of the overload because of their small size in relation to the large size of the line voltage connections, as shown in figure 10.3. The contacts will either be open or closed. If the contacts are open, the overload is bad (or there is an overload in the circuit). If the contacts are closed, the overload is good.

The line voltage part of the pilot duty overload indicates an overload by several methods: heat, current, and magnetism. All three methods determine the current flow, but they use different elements to determine an overload. The heat (thermal) type of pilot duty overload shown in figure 10.4(a) actually transfers current to heat. The current type of overload is similar in design to the magnetic. The current type of pilot duty overload, shown in figure 10.3, uses the current through a coil to indicate an overload and open

FIGURE 10.3. Pilot duty and line voltage connections of an overload. Photo courtesy of Texas Instruments Incorporated.

(a) Heat type of pilot duty overload   (b) Magnetic overload

*FIGURE 10.4.   Pilot duty overloads. (Arrows indicate pilot duty con-
nections.) Photos courtesy of Furnas Electric Co. and Heinemann
Electric Co.*

the pilot duty contacts. The magnetic overload shown in figure 10.4(b) uses
the strength of the magnetic field to open and close the pilot duty contacts.
The magnetic overload gives a certain amount of built-in time delay due to
its makeup.

To check the thermal element of a thermal overload, take a resistance
reading across the thermal element. If the resistance is above 0 ohms, the element is bad and the overload should be replaced. The magnetic and current
types of pilot duty overload can be checked in the same way. However, the
magnetic and current overloads use a coil that is larger than the element of a
thermal overload. This coil is connected in series with the load and therefore
indicates the current being used by the load. The coil in the magnetic or current types of pilot duty overload can be easily checked with an ohmmeter.
The resistance of the coil should be 0 ohms because it is part of the conductor going to the load. The overload should be replaced if the coil reads any
resistance.

Internal Overloads

Another type of overload is the internal overload used in hermetic compressors. This type of overload is actually embedded in the windings of the hermetic compressor motor, which gives it a faster response to overloads.

The early type of internal overload used separate terminals extending

FIGURE 10.5. Early type of internal overload of a hermetic compressor. Photo courtesy of Copeland Corporation.

from the inside of the compressor to the outside of the compressor terminal box, as shown in figure 10.5. This type of internal overload can be simply and easily checked with an ohmmeter to determine if it is open or closed. It is embedded in the winding but makes no electric connections to the windings.

The type of internal overload that is presently used is hard to check because it has no external connections. It is in series with the common terminal of the compressor motor. This type of overload must be checked as part of the windings of the motor, which makes it extremely hard to diagnose for troubles. The only way to check this type of overload is with an ohmmeter, just as you would check a motor. If an open is present, it could be due to the windings or the overload. Service personnel should never condemn a hermetic compressor that is temporarily overloaded. The compressor should be given ample time to cool.

Most semihermetic compressors have an internal thermostat embedded in the motor windings. The thermostat has a separate connection to the com-

FIGURE 10.6. Semihermetic overload terminals. Reproduced by permission of Carrier Corporation, © 1977 Carrier Corporation.

pressor terminals as shown in figure 10.6. An internal thermostat can easily be checked with an ohmmeter. If any resistance is read, the device is faulty and should be replaced.

10.3 THERMOSTATS

Some type of thermostat is used on almost all heating, cooling, and refrigeration equipment. Therefore it is essential to know how to correctly diagnose the condition of thermostats.

There are two basic types of thermostats used in the industry today: the line voltage thermostat and the low-voltage thermostat. The line voltage thermostat is used to make or break line voltage to a load. Its only function is to open or close a set of contacts on a temperature rise or fall. Thus the line voltage thermostat is usually simpler than the low-voltage thermostat because it does not have as many functions as the low-voltage thermostat. The low-voltage thermostat is used when a voltage lower than 110 volts— usually 24 volts—is used to operate a control system. The low-voltage thermostat can have many functions. It can stop and start a fan motor, operate a fan motor independently of other parts of the system, and do many other functions that are sometimes required in control systems. The line voltage thermostat is not as accurate as the low-voltage thermostat due to the contacts' larger size, which is necessary to carry the higher voltage. Low-voltage thermostats are almost always used on residential heating and cooling control systems and on many commercial and industrial systems. The line voltage thermostat is used on window air conditioners and commercial and industrial air-conditioning, heating, and refrigeration equipment.

Line Voltage Thermostat

The line voltage thermostat is usually easy to troubleshoot because of its simplicity. Most line voltage thermostats have two, three, or four terminals, as shown in figure 10.7. The most important element of checking line voltage thermostats is to be sure that the contacts are closed in the correct temperature range. Once it has been determined that the thermostat should be opened or closed, it can be checked with an ohmmeter. Or the control voltage can be checked at the equipment. The wires must be removed from the thermostat to check it with an ohmmeter. The ohmmeter will read 0 ohms if the thermostat is closed and infinity if it is open. The service mechanic must determine the correct terminals to check on the thermostat.

FIGURE 10.7. Line voltage thermostat.
White-Rodgers Div., Emerson Electric Co.

This information can be found on the unit's wiring diagram. In many cases it is difficult to remove the wires of a thermostat, so a voltage check is made.

Low-Voltage Thermostat

The low-voltage thermostat is more difficult to troubleshoot than the line voltage thermostat because of the many functions of the low-voltage thermostat. The low-voltage thermostat operates the heating and cooling of the system, operates the fan motor with the heating and cooling operations, operates the fan motor independently, often operates two-stage systems, operates damper motors, and operates a pilot function of a gas heating system. A common subbase of a low-voltage thermostat is shown in figure 10.8. All the letter designations, which indicate the different functions of the low-voltage thermostat, are also shown in the figure.

FIGURE 10.8. Common low-voltage thermostat subbase. Photo courtesy of Honeywell.

In troubleshooting a system the low-voltage thermostat and the subbase may be at fault. The low-voltage thermostat and subbase can be checked with an ohmmeter. This can be done at the equipment or at the junction point of the thermostat wires. Service mechanics seldom need to remove the thermostat and install a set of wires on it to check it. A low-voltage thermostat can also be checked by taking a voltage check at the equipment to ensure that the thermostat is functioning properly. The low-voltage thermostat and subbase are merely a point in the control system that is fed with low voltage. The thermostat sends on a voltage signal to the equipment, which must then operate to meet the conditions called for by the thermostat.

The chart shown in figure 10.9 can often be used in troubleshooting thermostats.

10.4 PRESSURE SWITCHES

Pressure switches are used on heating, cooling, and refrigeration systems to start or stop some electric load in the system when the pressure in the system dictates this action. Pressure switches are used as safety devices or as

CONDITION: (columns, left to right)
1. T/S jumpered; system won't work.
2. T/S jumpered; system works.
3. Room temp. overshoots T/S setting; too cold.
4. Room temp. doesn't reach setting; too warm.
5. System cycles too often.
6. System doesn't cycle often enough.
7. Room temp. swings excessively.

NOTE: T/S indicates room thermostat.

POSSIBLE CAUSES:

1	2	3	4	5	6	7	
●							T/S not at fault; check elsewhere.
		●					T/S wiring hole not plugged; drafts.
				●	●		T/S not exposed to circulating air.
		●	●				T/S not mounted level (mercury switch types).
		●	●				T/S not properly calibrated.
		●		●			T/S exposed to sun, source of heat.
	●				●		T/S contacts dirty.
	●		●				T/S set point too high.
		●					T/S set point too low.
	●		●				T/S damaged.
			●				T/S located too near cold air register.
	●						Break in T/S circuit.
		●	●		●	●	System sized improperly.

FIGURE 10.9.   *Troubleshooting chart for thermostats. Chart courtesy of Honeywell.*

operating controls. A pressure switch used as a safety device will stop an electric load when the pressure in a system reaches an unsafe condition. A pressure switch used as a safety device can be used to protect a refrigeration system from excessive discharge pressure or low suction pressure. It can be used in a gas heating system to protect the equipment from low or high gas pressure. It can be used to protect air-moving equipment from low air pressure.

An operating-control pressure switch is used to operate some load in the system. The most common use of an operating-control pressure switch is on a commercial refrigeration system to control the temperature of a walk-in cooler or freezer. Some pressure switches are also used to operate unloading devices on large compressors, to operate condenser fan motors to control the discharge pressure of a refrigeration system, and to operate pumps and cooling tower fans on water-cooled condensers.

The most important aspect of checking a pressure switch is to understand what it is used for in the system. The service technician must also determine if the pressure switch should be opened or closed. Once the service mechanic has determined what the pressure switch is used for and whether it should be opened or closed, it is easy to check the pressure switch. A resistance or voltage check should be made to determine if the pressure switch is opened or closed.

In troubleshooting pressure switches, it may be that the pressure switch is faulty or that the system is malfunctioning. The mechanic should take extra care to determine which condition exists.

The pressure switch can be stuck open or closed, it can open or close on

FIGURE 10.10. Buck-and-boost transformer. Photo courtesy of Acme Electric Corporation.

the wrong pressure, or there can be a mechanical problem with the switch itself. If a pressure switch is stuck open or closed, then the pressure in the system will have no effect on the pressure switch. The service mechanic will have no trouble detecting a faulty pressure switch that should be opened or closed and is keeping a load from operating as it should. The mechanic should make sure that the pressure switch is not stuck in a position that will result in damaged loads. If a pressure switch is not opening and closing on the right pressure, it should be correctly set or replaced.

On some occasions a pressure switch has been opened and closed so many times that the switch is worn out. Other mechanical linkage problems include broken springs, leaking bellows, corroded linkages, and broken linkages. Any condition that occurs in the mechanical linkage of a pressure switch will usually call for replacement of the pressure switch.

In most cases when a pressure switch opens, this is due to a malfunction in the system. But occasionally the pressure switch itself is at fault. The service mechanic must determine which condition is occurring. A pressure switch that is used as an operating control will often have to be reset but seldom replaced. The same troubleshooting procedure is used regardless of the function of the pressure switch.

10.5 TRANSFORMERS

A transformer is a device that is used to raise or lower the incoming voltage by induction to a more usable voltage for the control system. Some types of transformers are used to buck (lower) or boost (raise) the incoming voltage to an air-conditioning unit. A buck-and-boost transformer is used in conjunction with a voltage system that is too high or too low to supply the correct voltage to a system. Figure 10.10 shows a typical buck-and-boost transformer.

A transformer can be checked in two ways: a resistance check or a voltage check. An ohmmeter can be used to check the condition of the windings of a transformer. If the ohmmeter reads 0 ohms, the windings of the transformer are shorted. A reading of infinity indicates an open transformer. A measurable resistance indicates a good transformer. The number of ohms measured would be determined by the voltage of the transformer.

A transformer can also be checked by reading the output voltage if the correct primary voltage is applied. In some cases the transformer might check out as good if the load is not put across the secondary. But when the load is inserted in the line, the transformer voltage will not be enough to energize the load. In this case the transformer has a spot burnout and should be replaced. Often transformers are burned out because other devices in the circuit are being shorted. The service mechanic should take every precaution to prevent this from happening.

10.6 ELECTRIC MOTORS

Electric motors are the most common loads in any heating, cooling, or refrigeration system. They are used almost exclusively to cause the rotating motion of fans, compressors, pumps, and dampers. There are many different types of electric motors used in the industry. However, the type of motor used will have no effect on diagnosing the condition of the windings.

Open electric motors will fail in three different areas: the windings, the centrifugal switch, and the bearings. The windings can be checked with an ohmmeter for opens, shorts, and grounds. The condition of the centrifugal switch can best be determined by visual inspection after disassembling the motor. The bearings can usually be checked by turning the motor by hand to determine if there are hard or rough spots in the rotating movement of the motor.

The only part of a sealed motor that can be checked is the winding because there are no internal parts other than the bearings. However, there are several different types of starting apparatus that need to be checked. These components were discussed in chapter 7. Chapter 6 included the correct method of troubleshooting motors.

SUMMARY

Most heating, cooling, and refrigeration personnel are required to do some diagnosing of components. Thus it is imperative for personnel to understand how components work, to know how to use electric meters to check components, and to know the proper procedures for checking electric components. In this chapter we have presented some guidelines that personnel may find helpful in troubleshooting electric systems and their components.

Contactors and relays can be effectively diagnosed by checking the contacts, the coil, or the mechanical linkage.

There are several types of overloads that are used in the industry. The fuse can be checked with an ohmmeter. The circuit breaker can often be checked with a voltmeter. The line voltage overload can be checked with an ohmmeter. The pilot duty overload must have its contacts and controlling line voltage elements checked. These components can usually be checked with an ohmmeter. The internal overload used in hermetic compressors can best be checked with an ohmmeter.

Two basic types of thermostats are used in the industry. The line voltage thermostat can be checked with an ohmmeter. The low-voltage thermostat can also be checked with an ohmmeter but care must be taken with this type of thermostat because of its many functions.

There are two types of common pressure switches: low pressure and high pressure. Both can be serviced with either a resistance check or a voltage check. Transformers can also be checked by using either an ohmmeter or a voltmeter.

QUESTIONS

1. What is the first step in troubleshooting any system or component?

2. True or false. The best way to check the contacts of a contactor is by a visual inspection.

3. In troubleshooting contactors and relays, three areas must be checked: the _____, _____, and _____.

4. How do you make a voltage check to diagnose the condition of a set of contacts in a relay or contactor?

5. True or false. If the contacts of a contactor are out of alignment, the contactor must be replaced.

6. If the coil of a relay is shorted, the resistance reading will be _____ ohms.

7. What is the difference between an open coil and a shorted coil?

8. How do you check a fuse?

9. How do you check a circuit breaker?

10. When a line voltage overload is weak, an amperage check should be made to see what _____ is causing the overload to open.

11. What is the difference between a line voltage overload and a pilot duty overload?

12. True or false. The internal overload is difficult to check because it is in parallel to the common terminal of the compressor.

13. The service technician should never condemn a compressor with an internal overload until it has had time to _____.

14. The two types of thermostats used in the industry are _____ and _____.

15. Give some common applications of the low- and line voltage thermostats.

16. The line voltage thermostat usually has _____ function, while the low-voltage thermostat has _____ functions.

17. What do the letters R, Y, W, and G represent on a thermostat subbase?

18. How do you diagnose a bad pressure switch?

19. The most important aspect of checking a pressure switch is to understand its _____ in the system.

20. What is the proper procedure for checking a transformer?

11

Air-Conditioning Control Systems

INTRODUCTION

In the heating, cooling, and refrigeration industry there are many different types of control systems and they range from simple to very complex. The equipment used in residential air-conditioning units is usually simple and contains a limited number of components. Commercial and industrial air-conditioning equipment uses a more complex control system, with more emphasis on safety devices than is the case for residential control systems. The trend in the industry in the last few years has been to make residential air-conditioning control systems simpler but without any reduction in the safety. This has been done because of advances in control system design. At the present time the industry is using some solid-state control modules in control systems, but the control systems are by no means becoming completely solid state.

The control systems used in the industry today are combinations of different control systems for separate pieces of equipment, such as a furnace with an air-conditioning unit containing some method of interlocking. The control system has the same function whether the system is a single component or a combination of components. An air-conditioning control system is designed to operate a system automatically, incorporating all the necessary safety equipment by the use of a thermostat. The safety devices of a control system ensure that the loads of the system operate without any chance of damage. However, these safety devices are not always effective in preventing damage to the components because of the economy factor that must always play a part in the design of equipment. Beyond the safety devices of a control system are devices that control the operation of equipment to maintain a temperature at a certain range.

The installation mechanic of air-conditioning or heating equipment must be able to connect the control system of the equipment and the thermostats—and in many cases the furnace. Most manufacturers furnish a wiring diagram with equipment that shows the proper hookup of the equipment being used. In many cases, however, the installation mechanic must install equipment without a diagram. Therefore it is important for service personnel to know how to connect the controls correctly to ensure proper operation.

All heating, cooling, and refrigeration personnel are at one time or another required to be familiar with modern control systems. For example, salespeople may be called on to assist customers with control system design. Engineers are required to design control systems. Service mechanics are required to maintain and repair control systems.

About 85% of the problems in heating, cooling, and refrigeration control systems can be traced to some electrical problem. Although it is impossible to study all the different control systems used in the industry today, familiarity with the more common control systems will enable you to understand special control systems because of the similarities of design. In this chapter we will present the more common control systems in use in the industry today.

11.1 BASIC CONDENSING UNITS

A **condensing unit** is the portion of a split air-conditioning system that is mounted outside and contains the compressor, the condenser, the condenser fan motor, and the necessary devices to control these components. A split system is one that is divided into two parts, usually a condensing unit (outside) with a fan coil unit (inside). Figure 11.1 shows an air-cooled condensing unit.

In most cases the condensing unit is used with some type of equipment that will produce the air flow, and a coil must be mounted in the air flow. Almost all condensing units used in the industry today are air-cooled condensing units, which means that the condenser is cooled by air. Water-cooled condensing units may still be found in the field, but these units are rapidly being replaced with the smaller and more economical air-cooled condensing units.

When a condensing unit is used in an air-conditioning system, it is sepa-

FIGURE 11.1 Air-cooled condensing unit. Photo courtesy of Fedders Corp.

rate from the rest of the equipment. The control system must be connected to allow for complete control of the entire system.

The condensing unit takes the cool suction gas coming from the evaporator, compresses it, condenses it from a gas to a liquid, and forces it back to the evaporator. The standard components of an air-cooled condensing unit are the compressor, the condenser, the condenser fan motor, and all the devices used for control. The components of an air-cooled condensing unit are shown in figure 11.2. Most of the controls used to operate the condensing unit are mounted somewhere in the unit itself with the exception of the interconnection between the thermostat and the condensing unit.

FIGURE 11.2 Components of an air-cooled condensing unit. Photo courtesy of Fedders Corp.

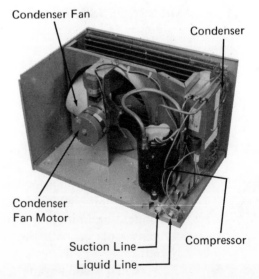

Condensing Unit Components for a Simple Control System

In early control systems all components necessary for operation of the total air-conditioning system were mounted in the condensing unit. This practice has all but disappeared in the industry because of the expense that it required and the realization that it was unnecessary. Most modern condensing units pick up their 24-volt power supply to operate the control system from the furnace transformer or from an indoor fan relay package. The evaporator fan motor usually is controlled by a relay when the system contains a furnace or fan coil unit and the condensing unit.

The compressor and condenser fan motor in a condensing unit are usually controlled by a contactor. The simplest control system used on condensing units today is a contactor that controls the operation of the entire condensing unit with the exception of the internal overloads in the compressor and condenser fan motor. With this type of control system, the contactor is energized whenever the thermostat is calling for cooling, even though one of the safety devices has the compressor or condenser fan motor cut out. This control system is the simplest and the least expensive of any condensing unit control system used today. A separate 24-volt source must be supplied from other components through the thermostat to the condensing unit.

All condensing units without a high-pressure switch must have an internal relief valve in the compressor that opens if the discharge pressure exceeds an unsafe level. This relief valve replaces the high-pressure switch on some units.

Figure 11.3 shows a control panel used on a condensing unit control system. Note the simplicity of the panel.

FIGURE 11.3. Control panel of a simple air-cooled condensing unit. Photo courtesy of Fedders Corp.

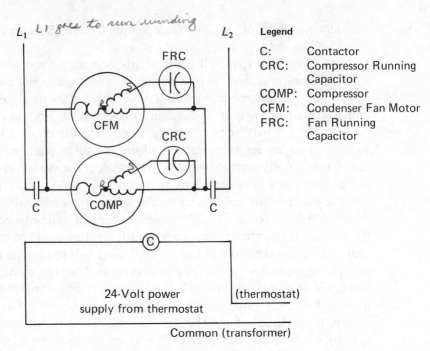

L₁ goes to run winding (handwritten)

CAPACITOR RUN OR START feeds start winding (handwritten)

FIGURE 11.4. *Schematic diagram of a simple air-cooled condensing unit with a minimum number of controls.*

Figure 11.4 shows the schematic diagram of this type of control system. The control system requires a supply of 24 volts to the contactor through the thermostat to operate the condensing unit. When 24 volts are applied to the contactor through the thermostat, the contactor will close, allowing 230 volts to be applied to the compressor and condenser fan motor, and they will start. This control system is easily diagnosed for trouble, with the exception of the overloads in the compressor and the condenser fan motors.

Condensing Unit Components for Complex Control Systems

Many manufacturers use a complex control system that contains certain safety devices to ensure safe operation of the condensing unit. The control system discussed in the preceding section had only the minimum required safety devices, which are often inadequate for a completely safe operation of the condensing unit. The more complex control systems incorporate a high-pressure and a low-pressure switch. Figure 11.5 shows the schematic diagram of a control system with the extra safety controls. If any of the safety controls open, the compressor and condenser fan motor stop.

Several manufacturers also use a device that protects the system from short cycling. This device maintains a certain period of time between the cycles of the equipment so that the system does not cut on and off in rapid succession (short cycling). This control system adds a relay and a time clock mechanism for this purpose. Other systems use a special relay to lock out the entire condensing unit until it can be reset manually.

There are many other control system designs that use more components for specific purposes. For example, large condensing units use a control relay or compressor relay to control a 240-volt contactor coil and the condenser fan motor. This arrangement is used because for a large contactor, 24 volts are less effective in pulling the contactor in than are 240 volts. There are many other designs, too numerous to discuss, used in the industry for the

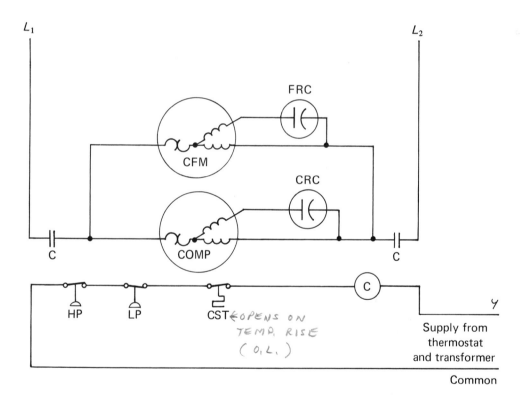

Legend

C:	Contactor	CRC:	Compressor Running Capacitor
COMP:	Compressor	HP:	High-Pressure Switch
CFM:	Condenser Fan Motor	LP:	Low-Pressure Switch
FRC:	Fan Running Capacitor	CST:	Compressor Safety Thermostat

FIGURE 11.5. Schematic diagram of a condensing unit with safety controls.

control of condensing units. In almost all cases, though, control systems have several basic similarities that make them fairly easy to understand.

Wiring

All condensing units come from the manufacturers with a wiring diagram and, in some cases, with an installation wiring diagram. Usually it is not difficult to follow the installation instructions and wire the condensing unit correctly. Figure 11.6 shows an installation diagram used to install a condensing unit to a furnace for an add-on air-conditioning system. This diagram shows the wiring hookup of a condensing unit and a gas furnace. It is used when installing a condensing unit and furnace.

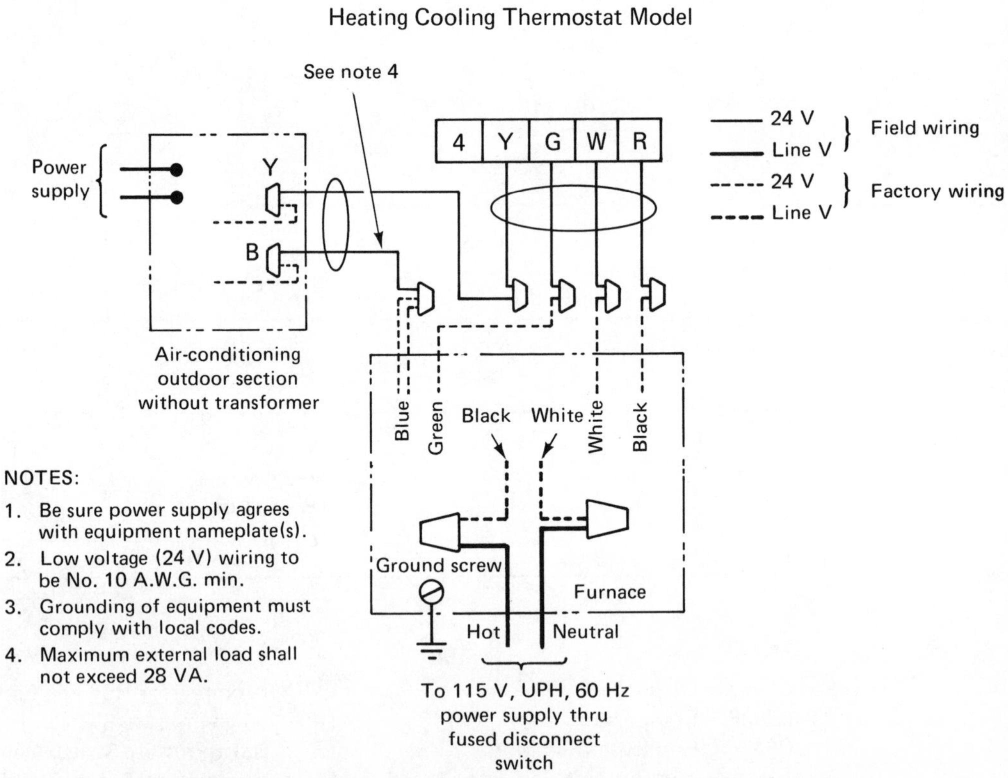

Heating Cooling Thermostat Model
See note 4
4
Y
G
W
R
24 V
Line V
Field wiring
24 V
Line V
Factory wiring
Power supply
Y
B
Air-conditioning outdoor section without transformer
Blue
Green
Black
White
White
Black
Ground screw
Furnace
Hot
Neutral
To 115 V, UPH, 60 Hz power supply thru fused disconnect switch
NOTES:
1. Be sure power supply agrees with equipment nameplate(s).
2. Low voltage (24 V) wiring to be No. 10 A.W.G. min.
3. Grounding of equipment must comply with local codes.
4. Maximum external load shall not exceed 28 VA.

FIGURE 11.6. Installation diagram for a condensing unit connected to a furnace. Diagram courtesy of General Electric Co., Central Air Conditioning Dept.

Troubleshooting

Any troubleshooting of the condensing unit can be done from the schematic because of the simplicity of most control systems. By using the schematic diagram, and having an understanding of the components, you should have no trouble with diagnosing a residential control system. However, you may have to study the more complex commercial and industrial control systems to diagnose problems in them.

11.2 PACKAGED UNITS

A **packaged air-conditioning unit** is built with all the components housed in one unit. In most cases packaged units are complete except for the power connections and the control connections.

Air-cooled packaged units are the most widely used types of packaged units. The units are used on applications ranging from small residential cooling units to large commercial and industrial air-cooled units. The smaller residential systems may be installed when the structure is built or at a later date.

An air-cooled packaged unit would usually be mounted outside the structure, as shown in figure 11.7. In some cases air-cooled packaged units can be installed inside with a remote condenser. The air-cooled packaged unit contains the compressor, evaporator fan motor, condenser fan motor if not of remote design, and all the necessary controls.

Water-cooled packaged units are usually used on commercial and industrial systems. The water-cooled packaged unit is shown in figure 11.8.

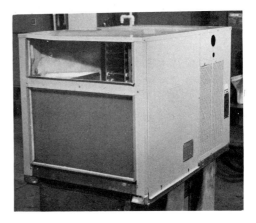

FIGURE 11.7.   Air-cooled packaged unit.
Lennox Industries, Inc.

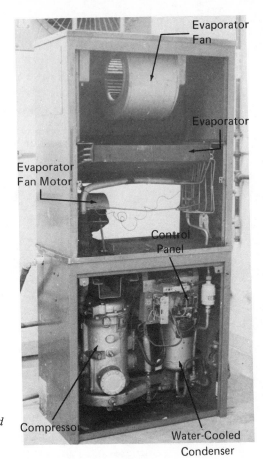

FIGURE 11.8. Water-cooled packaged unit. Photo courtesy of Airtemp Corp.

They are almost always installed when the structure is built. The water-cooled unit requires a cooling tower or some means of supplying water to the condenser. If a water-cooled unit is used, there must be interlocks to ensure that the cooling tower pump is operating when the compressor is operating. Water-cooled packaged units come complete with the compressor, evaporator fan motor, all the necessary controls, and the condenser.

Control System for Packaged Air-Conditioning Systems

Packaged unit controls are usually simple on small residential equipment and become more complex on the larger commercial and industrial equipment. The smaller air-cooled packaged units utilize a fairly simple control system with an interlock for the condenser fan motors. The large air-cooled and water-cooled units have a fairly complex control system. The large air-cooled packaged units are complex because of the zoning requirements that these units must meet in the rooftop applications in which they are usually used. Water-cooled packaged units utilize a line voltage control system with some type of interlock to start the cooling tower pump or other accessories that are required.

Air-cooled packaged units can incorporate some form of heating as well as cooling. The smaller packaged units usually have some easy means of installing electric resistance heat into the equipment. If gas heating is desired, a **gas pack** should be used, which is an electric air-conditioning and gas-heating system mounted in one unit. The larger air-cooled packaged units make electric and gas heat available to the consumer. Water-cooled packaged units and air-cooled packaged units usually use a hot-water coil, steam coil, or resistance heat for heating purposes.

Small Air-Cooled Packaged Units

The small air-cooled packaged units usually have a simple and easy-to-follow control system. The control system uses a contactor to operate the compressor and the condenser fan motor and an evaporator fan relay to control the evaporator fan motor. All necessary safety components are also included in the control system.

All manufacturers' control systems are not alike. Many use only the necessary safety controls, while others use more safety controls to ensure complete and safe operation.

Figure 11.9 shows the schematic diagram of a small air-cooled packaged unit and its controls. This unit would be used in a residence or small commercial application. The thermostat at the bottom of the schematic controls the operation of the unit. When the fan switch is in the "on" position or in the "auto" position, the thermostat calls for cooling, energizes the indoor fan relay, and, through its contacts, starts the indoor fan motor. The compressor is energized through a number of components. First the thermostat closes, energizing the control relay and closing its contacts. This energizes the outdoor fan motor and the contactor coil through the holding relay and the timer circuits. Once the contactor closes, the compressor will begin operation.

The installation of the air-cooled packaged unit is relatively simple because the only necessary connections are from the power source and the control source or thermostat. Most manufacturers include an installation diagram giving the correct connections and the control layout of the system.

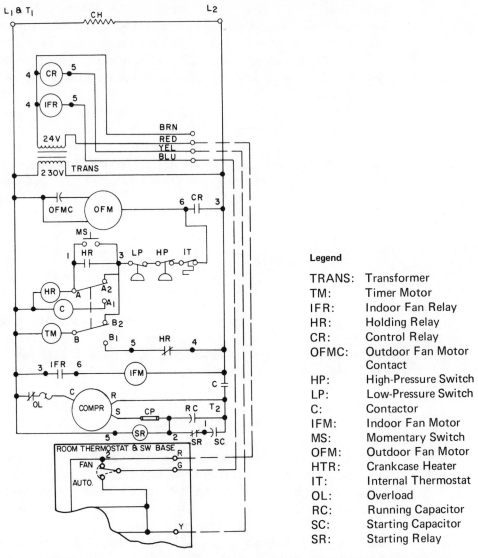

Legend

TRANS:	Transformer
TM:	Timer Motor
IFR:	Indoor Fan Relay
HR:	Holding Relay
CR:	Control Relay
OFMC:	Outdoor Fan Motor Contact
HP:	High-Pressure Switch
LP:	Low-Pressure Switch
C:	Contactor
IFM:	Indoor Fan Motor
MS:	Momentary Switch
OFM:	Outdoor Fan Motor
HTR:	Crankcase Heater
IT:	Internal Thermostat
OL:	Overload
RC:	Running Capacitor
SC:	Starting Capacitor
SR:	Starting Relay

FIGURE 11.9. Schematic diagram of a small air-cooled packaged unit used in a residence. Reproduced by permission of Carrier Corporation. © 1977 Carrier Corporation.

Gas-Electric Air Conditioners

The combination of electric air conditioning and gas heating in a packaged unit gives a system that is somewhat more complex than a straight air-cooled packaged unit. The main difference between the two units is all the extra components needed to combine a heating system with an air-conditioning system. The basic control system for the air conditioning is the same as the one for most small residential units. However, several components are added to take care of the heating control system.

Figure 11.10 shows a schematic diagram of the control system used on a modern gas-heating and electric air-conditioning packaged unit. The installation instructions should be followed when installing this type of system.

Referring to figure 11.10, the cooling mode of operation is energized through the thermostat to the contactor, which starts the compressor and outdoor fan motor. The indoor fan motor is started through the thermostat to the indoor fan relay. The heating mode of operation is a two-stage operation, with the first stage of the thermostat (heating) energizing the low fire and the second stage of the thermostat (heating) energizing the high fire if needed.

Air-Cooled Packaged Units with Remote Condensers and Water-Cooled Packaged Units

The air-cooled packaged unit with the remote condenser and the water-cooled packaged unit use a line voltage control system and are very similar in circuitry. These systems usually have line voltage control because they are shipped from the factory with all the controls mounted and wired, including the thermostat. These units are merely set in place when installed. The only necessary installation would be the power wiring connection, the condensate drain, and the piping.

The major difference between the water-cooled unit and the air-cooled unit is in the method the manufacturer uses to interlock the necessary components to cool the condenser. Figure 11.11 shows the schematic diagram of an air-cooled packaged unit with a remote condenser interlocked in the system. Figure 11.12 shows the schematic of a water-cooled packaged unit with a cooling tower pump interlock.

The unit shown in the schematic diagram of figure 11.11 has a line voltage control circuit. The switch marked "off," "fan," and "cool" operates the unit. With the switch set on "fan," the only portion of the unit that can operate is the indoor fan motor; it is energized through the manual switch. If cooling is desired, the switch must be set on "cool," which will not interrupt the operation of the indoor fan motor. This circuit energizes the outdoor fan motor through the outdoor fan relay. The compressor is energized through the contacts of the contactor, which is energized through the holding relay and the timer.

The unit shown in the schematic diagram of figure 11.12, a small water-cooled packaged unit, also has a line voltage control circuit. The switch

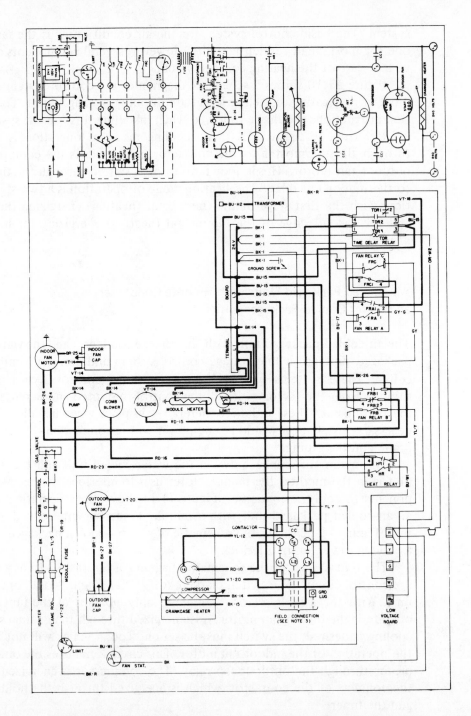

FIGURE 11.10. Schematic diagram of an electric air-conditioning and gas-heating unit. Diagram courtesy of Amana Refrigeration, Inc.

SCHEMATIC

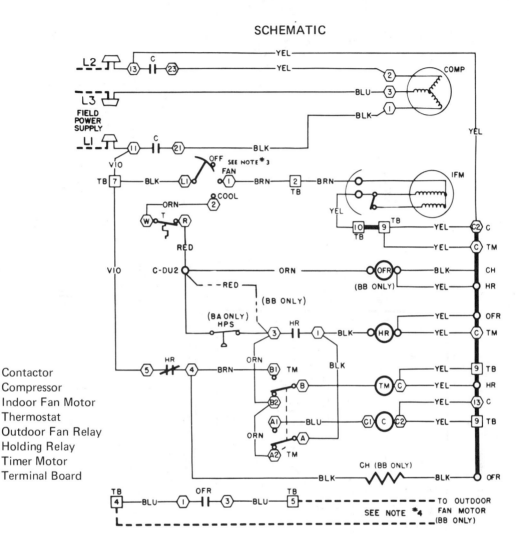

FIGURE 11.11. Schematic diagram of an air-cooled packaged unit with a remote condenser. Reproduced by permission of Carrier Corporation. © 1977 Carrier Corporation.

Legend

C: Contactor
COMP: Compressor
IFM: Indoor Fan Motor
T: Thermostat
OFR: Outdoor Fan Relay
HR: Holding Relay
TM: Timer Motor
TB: Terminal Board

marked "fan" and "cool" actually stops and starts the unit. When the switch is in the "fan" position, the fan will run. When the switch is moved to "cool," the fan will continue to run and the cooling thermostat will operate the compressor through the contactor.

Installation of these air conditioners is fairly simple because no connections are necessary, with the exception of the power wires. However, water connections are necessary on the water-cooled units and refrigerant lines are needed between the outdoor remote condenser and the inside packaged unit.

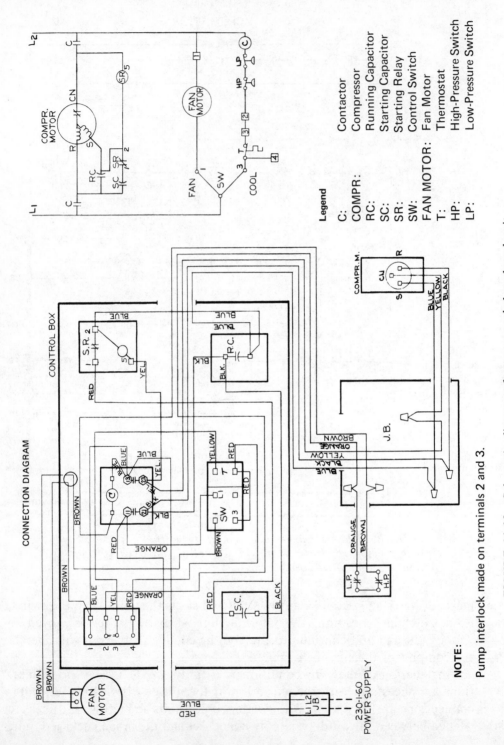

Legend

C: Contactor
COMPR.:   Compressor
RC: Running Capacitor
SC: Starting Capacitor
SR: Starting Relay
SW: Control Switch
FAN MOTOR: Fan Motor
T: Thermostat
HP: High-Pressure Switch
LP: Low-Pressure Switch

NOTE:

Pump interlock made on terminals 2 and 3.

FIGURE 11.12.   Schematic diagram of a water-cooled packaged unit. Reproduced by permission of Carrier Corporation. © 1977 Carrier Corporation.

Rooftop Units

The air-cooled packaged units used in rooftop applications are complex because they usually have some type of heating to go along with the air conditioning and, in some cases, some type of zone control. The air-conditioning and heating control system is usually simple. However, when it is connected to all the zone controls, the system becomes difficult to service and install.

As their name implies, these units are usually installed on the roof of a structure. But they could just as easily be installed on ground level when the necessary changes are made in the duct work connections. Very often the electric connections of these systems are hard to install because of the number of wires required for the control circuit and the power wiring. The installation instructions furnished with these units are usually very well written and give explicit instructions.

11.3   FIELD WIRING

In all heating, cooling, and refrigeration electric systems there is a certain amount of wiring that must be connected to the equipment after it has been set in place. This wiring must be installed by the installation mechanic and is called **field wiring**.

The **factory-installed wiring** is the wiring that is installed at the factory. It usually takes care of the connections between the components in the control panel and the system components. The factory wiring has been sized, color-coded, and installed in the control system to operate the equipment properly. The remainder of the wiring, whether it be the power wiring or control wiring, must be connected in the field.

Power Wiring

The power wiring of an air-conditioning and heating system is usually simple and easy to install on residential systems. There are two power connections that must be made on a split air-conditioning and heating system: the connections to the condensing unit and the connections to the evaporator fan motor or furnace. Figure 11.13 shows the power wiring on a condensing unit and a furnace with the correct connections from the distribution panel in the residence.

The power connections for a residential air-cooled packaged unit are

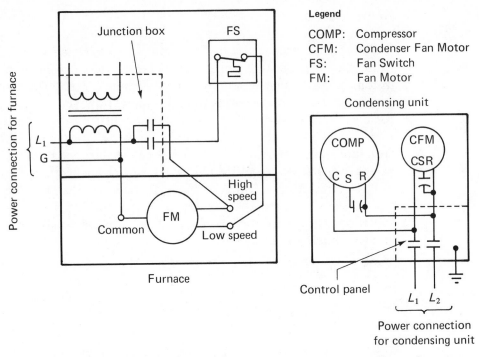

FIGURE 11.13.   *Power wiring for a condensing unit and a furnace (no control wiring is shown).*

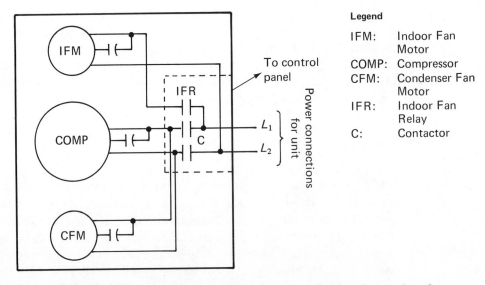

FIGURE 11.14.   *Power connections for a small air-cooled packaged unit (no control wiring is shown).*

shown in figure 11.14. This circuit is relatively simple, containing only the power connections and some type of bonding ground.

The electric supply wiring of a commercial and industrial system is somewhat more complex than the simple residential systems. The use of three-phase current in commercial and industrial structures does not add to the complexity of the system, only to the number of power wires that must be supplied to the system. The evaporator fans on any split commercial and industrial system would be supplied with a separate power source. The condensing unit would be supplied with its own power source. Figure 11.15 shows the supply wiring that must be installed with a fan coil unit that uses three-phase current.

Heating Systems. Commercial and industrial systems use various types of heating systems. In a system that uses electric resistance heat, a large power supply would be required to operate the heat. If some type of gas heating is used, the wiring would not have to be as large as that used with electric heat. Figure 11.16 shows the supply wiring hookup of a condensing unit, fan coil unit, and an electric duct heater. In many cases hot water or steam is used, but there are no additional power connections that must be made except for the pumps.

Large rooftop or packaged units equipped with gas heating usually require only one power source to the entire system. The rooftop units that incorporate electric resistance heat are usually supplied by two or more power sources because of the large load created by the electric heat. One of the power sources feeds the air-conditioning components and the evaporator fan motor; the other source feeds the electric heat.

FIGURE 11.15. Power wiring for a three-phase fan coil unit (no control wiring is shown).

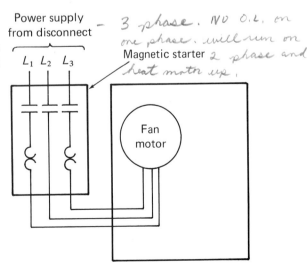

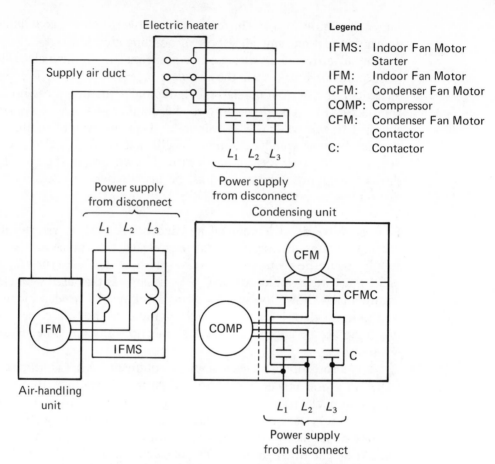

FIGURE 11.16. Supply power wiring for a condensing unit, fan coil unit, and large electric duct heater (no control wiring is shown).

There are many types and manufacturers of equipment in the industry today and it is impossible to show all the supply power connections in this section. However, it is important that the installation mechanic study other methods of connecting the power wiring to modern equipment.

Sizing Wires and Fuses. When installing heating, cooling, and refrigeration systems, air-conditioning personnel in charge of the installation must make sure that all power wiring is of the correct size and type. The distance that the power wiring must be run is very important because of the voltage drop that can occur on long circuit runs. The installation instructions usually give a wiring chart with the length that is allowable for each size of wire. Installation personnel should be capable of sizing the wire if it is not listed in the installation instructions (see chapter 5).

It is also important to follow the manufacturer's recommendations for the fuse or breaker size. Remember: Before any heating, cooling, or refrigeration system can be expected to operate properly, it must first be supplied with the correct size of wire and fuse to deliver the proper voltage to the system.

Control Wiring

The control wiring of heating and air-conditioning systems is just as important as the supply power wiring. In most cases low-voltage control systems are used on residential and small commercial applications. On large commercial and industrial applications, line voltage control systems are often used. Many large commercial and industrial systems are controlled by special control systems. These specialized control systems can be electric, pneumatic, or electronic. Specialized control systems are a complete subject in themselves and hence will not be covered in this section.

Residential.   All residential packaged air conditioners use a low-voltage control system that connects the thermostat to the unit, as shown in figure 11.17. These control systems are simple and easy to install by following the installation instructions.

Residential split systems are usually easy to install. They all incorporate a 24-volt control system, with the low-voltage power being supplied from an indoor fan relay package, the furnace, or the condensing units. Figure 11.18 shows one type of indoor fan relay package that is in common use in the industry today. The indoor fan relay package acts as a junction point between the thermostat, the furnace, and the condensing unit. The package has a terminal board that is labeled with easily identifiable letter identifications.

The control system of a residential split system can be connected several different ways, and the manufacturer's installation instructions should be followed. Some manufacturers use a transformer mounted in the condensing unit that supplies the low-voltage to the system, but this practice is becoming unpopular because of the advancements made in the indoor fan relay packages. In some cases the furnace transformer is used as the low-voltage power supply, but caution should be taken to ensure that the proper size is used. The more modern condensing units are usually designed without the low-voltage transformer but with two low-voltage connections instead. Figure 11.19 shows a control layout for a furnace, indoor fan relay package, thermostat, and the condensing unit. Figure 11.20 shows a control system of a furnace and a condensing unit with the thermostat and the low-voltage power supply coming from the furnace transformer.

251

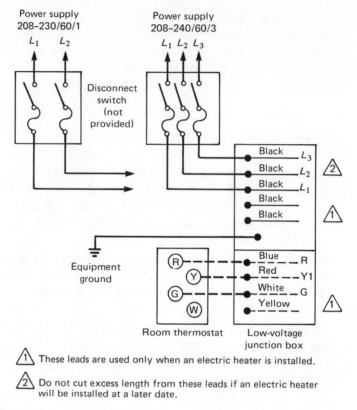

FIGURE 11.17. Control connections for a residential packaged unit. Diagram from Lennox Air Conditioning Company.

Industrial. Control systems for relatively small commercial units are similar in design to the residential low-voltage system. The only exception is that the system uses a control relay that supplies the condensing unit with 230 volts while using a normal 24-volt control system. Two-stage heating or two-stage cooling can also be used with a low-voltage control system.

Sizing Wire. Control wiring in most cases is a small wire (No. 18 to No. 20) that is usually covered by a rubber jacket when using a low-voltage control system. Thermostat wire can be purchased single-stranded or multiple-stranded, with or without a rubber jacket protecting the small thermostat wires. The size of the control wiring should follow the manufacturer's specifications.

The wiring of air-conditioning and heating systems is one of the most important factors in the installation of equipment. The control system is actually the heart of the electric system because its function is to properly con-

FIGURE 11.18. A common indoor fan relay package. Photo courtesy of Essex Group, Controls Div.

trol the entire system. The life of the equipment can be drastically cut by undersized wire or by having loose connections. Therefore, wiring to all components must be sized correctly. The installation instructions are the best place to look for the proper method of wiring and the proper connection points. Overlooking the smallest detail in the electric connections of a system can give trouble and improper operation.

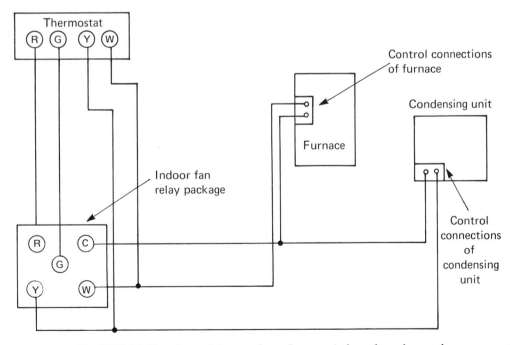

FIGURE 11.19. Control layout for a furnace, indoor fan relay package, thermostat, and condensing unit. Reproduced by permission of Carrier Corporation. © 1977 Carrier Corporation.

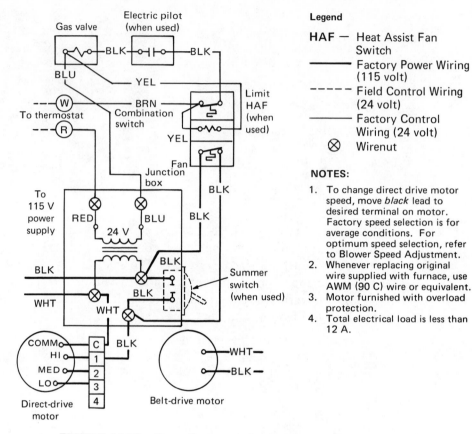

FIGURE 11.20. *Control system of a furnace and a condensing unit with the thermostat and low-voltage power supplied from the furnace. Reproduced by permission of Carrier Corporation. © 1977 Carrier Corporation.*

Legend

HAF — Heat Assist Fan Switch
———— Factory Power Wiring (115 volt)
– – – – Field Control Wiring (24 volt)
———— Factory Control Wiring (24 volt)
⊗ Wirenut

NOTES:

1. To change direct drive motor speed, move *black* lead to desired terminal on motor. Factory speed selection is for average conditions. For optimum speed selection, refer to Blower Speed Adjustment.
2. Whenever replacing original wire supplied with furnace, use AWM (90 C) wire or equivalent.
3. Motor furnished with overload protection.
4. Total electrical load is less than 12 A.

SUMMARY

Air conditioning, heating, and refrigeration control systems range from simple control systems with a simple line voltage thermostat to very complex systems using many different devices to control the temperature of each zone in a large structure. It is understandable that a large air-conditioning unit with a very expensive compressor will require a larger and better control system than the smaller residential units.

The electric system of an air-conditioning unit covers the power wiring as well as the control wiring, and in most cases they should be treated as one.

The electric system of a small condensing unit used in the residential or small commercial area is designed with only the bare necessities in controls.

Most of these smaller units are designed with a contactor and few safety controls in the condensing unit. The 24-volt supply must come from other equipment. The compressors of these units almost always contain an internal overload and this should be taken into consideration when working with these units. Some of the smaller commercial systems will have various added features to prevent the equipment from short cycling and to provide other important advantages.

Packaged air-conditioning equipment can take two different forms. The small residential and commercial packaged units are completely self contained and must have an outside air source. The straight commercial packaged units are built in one piece and are used in many cases in a free-standing position within a structure; they may be water-cooled or air-cooled with a remote condenser.

The packaged units that are mounted outside come in various designs. They range from a 2-ton air conditioner for a residence to a 50-ton or larger rooftop unit used in commercial buildings.

The straight packaged units usually use a low-voltage control system. Some of the larger units will have a line voltage control system within the unit. The free-standing packaged units almost always have a line voltage control system because they are prewired at the factory in this manner. All packaged units are prewired at the factory. In the straight packaged units the control wiring must be field installed. The free-standing packaged units are completely wired at the factory.

All heating, cooling, and refrigeration equipment will require some kind of electric connection in the field, if nothing more than the power connections. Most equipment will require a control circuit and a power circuit before the equipment will operate. The power circuit consists of the electric source carried to the unit. In some cases the evaporator fan motor will require a separate source. Control wiring is required between the thermostat and the equipment, or two sections of equipment, to make one complete system. The manufacturer almost always supplies an installation diagram with the equipment. This diagram will give wire and fuse sizes that are recommended by the manufacturer and should be followed.

QUESTIONS

1. What is the purpose of an air-conditioning control system?

2. What is the major difference between the controls on a small residential condensing

unit and a large commercial condensing unit?

3. What is a condensing unit?

4. Briefly describe how a condensing unit operates.

5. Most condensing units pick up the 24-volt power supply to operate the control system from _____ or _____.

6. True or false. The simplest control system used on condensing units is a contactor.

7. An internal relief valve is used in a compressor in place of a _____.

8. Explain the operation of a device that prevents short cycling in a system.

9. What is a split air-conditioning system?

10. Draw a diagram of the wiring of a simple residential air-cooled condensing unit used with a furnace, having a contactor, condenser fan motor, and a compressor.

11. What is the usual control voltage of a residential heating and cooling system and from where does it receive it?

12. How would a low-pressure and a high-pressure switch be connected into the electric system of a residential system?

13. What would be the reason for using a 240-volt control system in a large condensing unit?

14. What is a packaged air-conditioning unit?

15. An air-cooled packaged unit is usually mounted _____.

16. Why is it necessary to interlock the cooling tower pump and fan motor into the control circuit of a water-cooled packaged unit?

17. A gas pack is _____.

18. Line voltage control systems are used with what types of packaged units?

19. What is the major difference between the water-cooled unit and the air-cooled unit?

20. What is field wiring?

21. Name the two types of power connections that must be made on a split air-conditioning and heating system.

22. True or false. Commercial and industrial systems usually use electric resistance heat.

23. What are two major considerations an installation mechanic must take into account when installing air-conditioning power wiring?

24. What type of control system is used on most residential and small commercial equipment?

25. What is the purpose of an indoor fan relay package?

26. Control wiring in most cases is _____.

27. The heart of the electric system in heating, cooling, and refrigeration equipment is _____.

28. What type of electric connections must be made to a residential packaged unit by the installation mechanic to ensure proper operation?

29. What type of electric connections must be made to a commercial and industrial packaged unit by the installation mechanic to ensure proper operation?

12

Control Systems: Circuitry and Troubleshooting

INTRODUCTION

Control systems used in the heating, cooling, and refrigeration industry are designed to control the electric loads of a system to maintain the desired temperature of a given area or medium. Control systems must contain the necessary safety components to ensure safe operation of the equipment.

There are many different types of control systems used in the industry today, but most systems have certain components and circuits in common, depending on the control voltage and the type of equipment. Low-voltage control circuits are different from line voltage control circuits because of the devices that can be used with each type of system. The basic control system contains several components that are common to all control systems, such as the compressor, the evaporator fan motor, and the condenser fan motor, along with their controlling devices.

It is essential that installation and service mechanics understand the operation of the basic circuits of control systems to be able to install or troubleshoot the smaller air-conditioning and heating systems that are common to the industry. All equipment will not have the same type of controls and may use different methods of controlling certain devices. In this chapter we will be looking at basic control system circuits and procedures for troubleshooting control systems.

12.1 BASIC CONTROL CIRCUITS

All control systems have certain circuits and components in common. Most air-conditioning systems have a contactor to start and stop the compressor. The compressor is the largest electric load in an air-conditioning system. Therefore, it is necessary to use a contactor or starter to control its ampacity.

The control of the condenser fan motor is common on many systems, but it does vary due to the new high-energy equipment that is being produced. Circuits controlling the evaporator fan motor are common because of the use of an evaporator fan relay, but on heating systems different methods are used. The safety devices used on control systems vary greatly, but they are usually connected into the control circuit in series.

Now let us look at these control system components in detail.

Compressor Control Circuits

The first basic circuit of almost any residential or small commercial air-conditioning control system is the device that starts and stops the compressor. On a low-voltage control system, the thermostat opens or closes to energize or deenergize the contactor, which starts or stops the compressor, as shown in figure 12.1. This circuit is common to almost all residential and small commercial systems in operation today.

The condenser fan motor is usually connected directly to the contactor to ensure that it is operating whenever the compressor is in operation, as shown in figure 12.1.

Evaporator Fan Motor Control Circuits

All evaporator fan motor circuits are almost alike in the smaller ranges of equipment. The evaporator fan motor operates from a relay controlled by the thermostat. If the thermostat fan switch is set on the "on" position, the indoor fan relay will be energized by the cooling function of the thermostat. Figure 12.2 shows the indoor fan relay with the connections of the fan motor and the thermostat. Notice the fan switch on the thermostat and note how the indoor fan is cut on or off by use of the indoor fan relay. This control system is common only to air-conditioning systems. On heating installations other controls are needed to start the fan.

Some very small packaged units use a unique control system for the evaporator fan motor, as shown in figure 12.3. This type of control of the

Legend

C: Contactor
COMP: Compressor
CFM: Condenser Fan Motor
CRC: Compressor Running Capacitor
CFMC: Condenser Fan Motor Capacitor
TR: Transformer

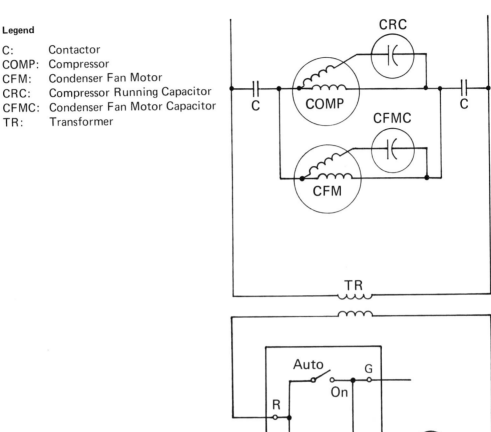

FIGURE 12.1. Schematic diagram of a contactor circuit controlling a compressor, showing the connection of the condenser fan motor.

evaporator fan motor is used because there is only one fan motor to operate, which works as both an evaporator fan motor and a condenser fan motor. In this case it is imperative that the fan motor operate whenever the compressor operates in order to cool the condenser. When the fan is to be energized on the thermostat through the fan switch for continuous operation, the fan is energized through an indoor fan relay. When the fan is energized through the cooling sections of the thermostat, the fan is controlled by the contactor. Most evaporator fan motor circuits are controlled by an indoor fan relay. Only in rare instances will the evaporator fan motor be energized by some other means.

Legend

COMP: Compressor
C: Contactor
CRC: Compressor Running Capacitor
CFM: Condenser Fan Motor
CFMC: Condenser Fan Motor Capacitor
IFR: Indoor Fan Relay
IFM: Indoor Fan Motor
TR: Transformer
CT: Cooling Thermostat

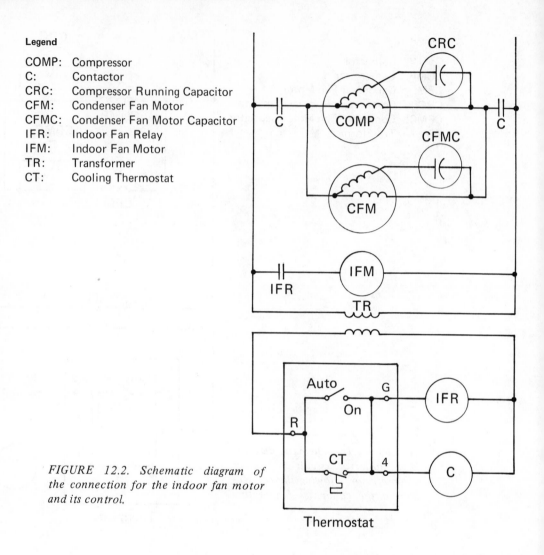

FIGURE 12.2. Schematic diagram of the connection for the indoor fan motor and its control.

Condenser Fan Motor Control Circuits

Almost all condenser fan motors on residential air-conditioning systems are controlled by the contactor because of the simplicity and the economy of using one control. Some condenser fan motors are cycled on and off through an outdoor thermostat set at a temperature that will maintain a constant head pressure. This allows better system efficiency. This type of control is accomplished on residential and small commercial systems by inserting the thermostat in series with the condenser fan motor. Other methods are used in the larger commercial and industrial applications.

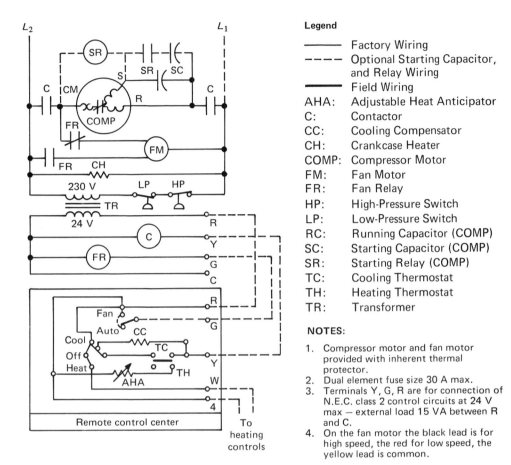

Legend

———— Factory Wiring
- - - - Optional Starting Capacitor, and Relay Wiring
▬▬▬▬ Field Wiring

AHA:	Adjustable Heat Anticipator
C:	Contactor
CC:	Cooling Compensator
CH:	Crankcase Heater
COMP:	Compressor Motor
FM:	Fan Motor
FR:	Fan Relay
HP:	High-Pressure Switch
LP:	Low-Pressure Switch
RC:	Running Capacitor (COMP)
SC:	Starting Capacitor (COMP)
SR:	Starting Relay (COMP)
TC:	Cooling Thermostat
TH:	Heating Thermostat
TR:	Transformer

NOTES:

1. Compressor motor and fan motor provided with inherent thermal protector.
2. Dual element fuse size 30 A max.
3. Terminals Y, G, R are for connection of N.E.C. class 2 control circuits at 24 V max — external load 15 VA between R and C.
4. On the fan motor the black lead is for high speed, the red for low speed, the yellow lead is common.

FIGURE 12.3. Schematic diagram of a small packaged unit with one fan motor. Reproduced by permission of Carrier Corporation. © 1977 Carrier Corporation.

Safety Control Circuits

Almost all safety controls used in residential and small commercial systems are connected in series with the contactor coil. The only exception is the internal thermostat in a compressor, which breaks the power wires inside the compressor. Any other safety devices used in the control system, such as a high-pressure switch or a low-pressure switch, are connected in series with the contactor coil. The series connection ensures that in the case of an unsafe condition, the compressor will be deenergized.

Figure 12.4 shows a high-pressure and a low-pressure switch in the control circuit. Any additional safety devices would be connected in the same manner. The internal overload in the compressor is connected in series with

Legend

C: Contactor
COMP: Compressor
CRC: Compressor Running Capacitor
CFM: Condenser Fan Motor
CFMC: Condenser Fan Motor Capacitor
IFR: Indoor Fan Relay
IFM: Indoor Fan Motor
HP: High-Pressure Switch
LP: Low-Pressure Switch
CH: Crankcase Heater

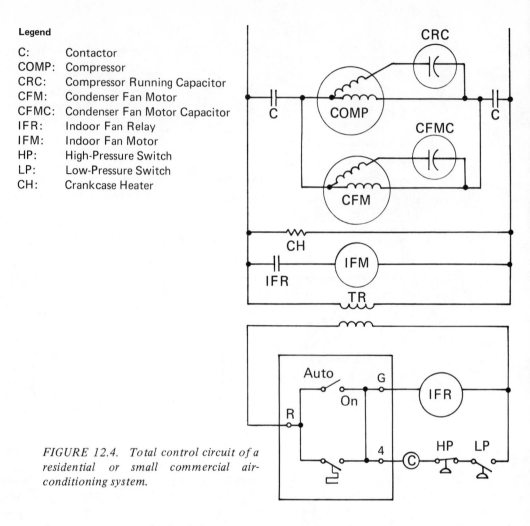

FIGURE 12.4. Total control circuit of a residential or small commercial air-conditioning system.

the common terminal of the compressor so that the compressor will be de-energized when an overload occurs.

12.2 TOTAL CONTROL SYSTEM OF RESIDENTIAL UNITS

The total control circuit of a residential or small commercial air-conditioning system is shown in figure 12.4. Each circuit can be discussed separately as long as each part in the circuit is complete in the total operation of the control system.

The thermostat is a switching point of the low-voltage supply to the necessary components. Figure 12.5 shows the schematic of the thermostat.

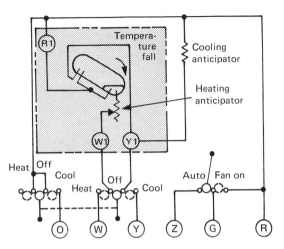

FIGURE 12.5. Schematic of a thermostat for a total residential control system. Diagram courtesy of Honeywell.

From the schematic it can be seen that the indoor fan motor can operate continuously or cycle with the cooling thermostat, depending on the position of the fan function switch. The remainder of the thermostat operates the heating or cooling or both. The cooling thermostat will close or open to the correct temperature, as well as control the heating by the heating thermostat.

Figure 12.3 shows that the compressor and in most cases the condenser fan motor operate when the contactor is energized. The contactor is energized by the cooling thermostat until the temperature is decreased to the desired range. The evaporator fan motor is energized on cooling by the indoor fan relay and on the heating cycle by a fan switch contained in the furnace.

As can be seen in figure 12.4, the safety devices are connected in series with the contactor coil, with the exception of the internal overload in the compressor, which is in series with the compressor common terminal. In some cases manufacturers use a crankcase heater that is energized at all times to keep liquid refrigerant from seeping into the compressor crankcase. This control system is commonly used on residential and small commercial air-conditioning and heating systems.

12.3 ADVANCED CONTROL SYSTEMS

All types and designs of control systems are used in commercial and industrial air-conditioning, heating, or refrigeration systems. The control systems used in equipment of from 7.5 to 100 tons cannot afford to be as simple and use as few components as the smaller systems. This type of control system must ensure that the electric loads are operating in a safe manner because of their greater cost. The control of this type of equipment must also be more complex than the smaller systems because of its size and operation.

Most large systems use some method of capacity control because of the variations in the load in the conditioned area. When capacity control is used on these larger air-conditioning systems, it usually involves a more complex control system. Also, most large air-conditioning and heating systems have several important control circuits that prevent short cycling and operation of the equipment under conditions that would cause damage. Thus we see that large equipment uses many more components than small equipment because of size, cost, capacity control, and general design of the larger systems.

Compressor Motor Controls

It would be impossible to cover all the control circuits used in the large equipment. However, the more common control circuits will be covered in detail.

There are two methods of energizing the compressor on the large systems, part winding and across the line. The contactor or magnetic starter is used to energize the compressor on all systems. An across-the-line type of motor starting requires only a single contactor, as shown in figure 12.6(a). However, a part winding motor would require two contactors along with a time-delay relay. The part winding motor is used to allow the compressor to start easier, with less wear and tear. A part winding motor hookup is shown in figure 12.6(b). The part winding motor is actually broken down into two separate windings, with the first winding energized 1.5 to 3 seconds before the second winding is energized. This increases the motor's starting efficiency.

Water Chiller Control Systems

Water chillers are refrigeration systems that cool the water pumped into other parts of the system to maintain the desired condition of a specific area. The control on a water chiller will usually come from the chilled water temperature, which will relate the information back to the refrigeration control system. This type of control can also include a pump-down system. The system maintains a certain chilled water temperature at all times. A typical control system for a water chiller is shown in figure 12.7.

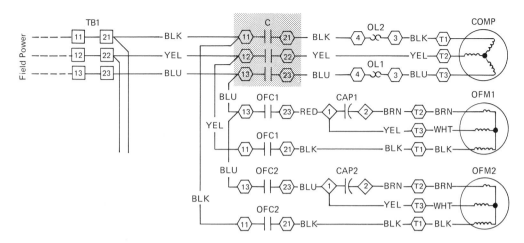

(a) Across-the-line starter (shaded area)

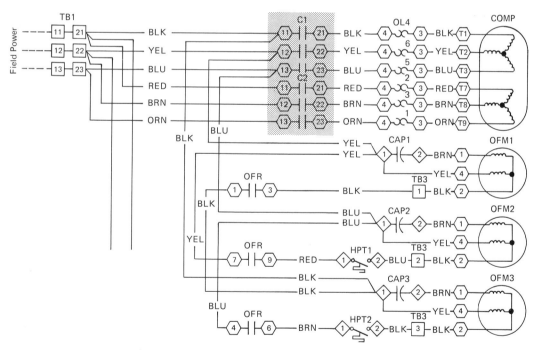

(b) Part winding motor hookup (shaded area)

FIGURE 12.6. Schematic of a part winding motor hookup (shaded areas). Reproduced by permission of Carrier Corporation. © 1977 Carrier Corporation.

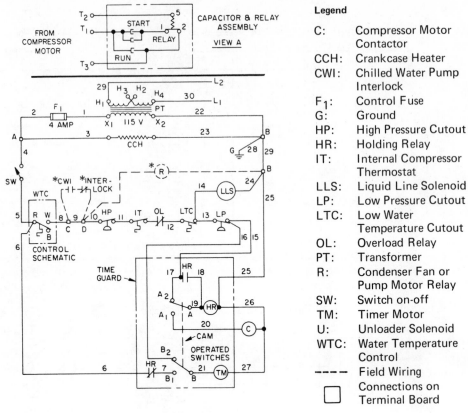

Legend

C: Compressor Motor Contactor
CCH: Crankcase Heater
CWI: Chilled Water Pump Interlock
F_1: Control Fuse
G: Ground
HP: High Pressure Cutout
HR: Holding Relay
IT: Internal Compressor Thermostat
LLS: Liquid Line Solenoid
LP: Low Pressure Cutout
LTC: Low Water Temperature Cutout
OL: Overload Relay
PT: Transformer
R: Condenser Fan or Pump Motor Relay
SW: Switch on-off
TM: Timer Motor
U: Unloader Solenoid
WTC: Water Temperature Control
- - - - Field Wiring
☐   Connections on Terminal Board

NOTES:

1. Interlock between terminals C and D: Condenser Water Pump or Fan Motor Overloads.

2. Remove wire number 9 between terminals C and D when interlock is added.

FIGURE 12.7. Typical control system for a water chiller. Reproduced by permission of Carrier Corporation. © 1977 Carrier Corporation.

Evaporator Fan Motor Controls

The control of the evaporator fan motor in large commercial and industrial air-conditioning or heating systems comes from a source separate from the condensing unit. Fan motors are generally designed to operate at all times in the larger systems and are cut on and off by a clock or a system switch. These fan motors are interlocked into the control system of the condensing unit to ensure the operation of the fan motor whenever the condensing unit is operating.

The smaller commercial air-conditioning and heating units usually cut off their fan motors on the "off" cycle of the equipment. However, they are equipped with a fan switch that allows continuous operation.

There are various other methods used to control evaporator fan motors, but none is as common as the constant air flow and the fan cycling with equipment operation.

Condenser Fan Motor Controls

The condenser fan motor used on most commercial and industrial air-conditioning units is usually designed to maintain a constant head pressure by operating a certain number of condenser fans when the outside temperature drops below 75°F. This is accomplished by using thermostats to operate the condenser fan motors at certain outdoor temperatures or by using high-pressure switches to cycle the fans only when they are needed. The condensing temperature of a condensing unit should be maintained at or around 90°F for proper operation of the air-conditioning equipment.

Water-Cooled Condenser Interlocks

Water-cooled condensing units require an interlock in the condensing unit control circuit. The interlock ensures that the air-conditioning unit does not start unless the cooling tower pump and the cooling tower fan motor are operating, as shown in figure 12.7. If the air-conditioning equipment were to operate without the pump operating, there could be damage to the compressor because of high head pressure.

12.4 METHODS OF CONTROL ON ADVANCED SYSTEMS

There are many methods of controlling commercial and industrial systems. The total control system is often totally different from the control system of each piece of equipment, with the exception of the initiation and termination of the operation of the equipment.

The control initiation is different from the low-voltage control system because of the size of the contactors and the additional components. These systems sometimes use a low-voltage control system, but it would only be used to energize the main control system.

Most large commercial and industrial systems have a main control system that maintains the control of the total structure. It only initiates and terminates the operation of the equipment to deliver the heating and cooling medium at a desired temperature. The total control system takes care of the remaining conditions. Therefore, the air-conditioning control is accomplished by starting or stopping the equipment. The control system of the equipment takes care of the operation of the equipment.

Most systems are equipped with some method of capacity control, which allows the equipment to operate at from 100% to 33% of its capacity. The smaller commercial and industrial systems usually operate with a standard low-voltage staging thermostat that is capable of capacity control to some extent.

In this section we will discuss some basic control methods used on the larger commercial and industrial systems.

Small Commercial and Industrial Systems

The small commercial and industrial systems often use a simple low-voltage control system to maintain the desired temperature in the structure, with some staging accomplished by the thermostat. However, in many cases low-voltage cannot be produced in sufficient strength to adequately close all the contactors or relays in a system. In this case the low voltage would control a relay to supply 110 volts or 220 volts to the cooling equipment.

Figure 12.8 shows a common control system of a 5-ton split air-conditioning system with a fan coil unit. A control relay starts and stops the operation of the condensing unit by supplying line voltage to the unit. This control relay is operated by a thermostat. The thermostat also operates the evaporator fan motor by energizing a magnetic starter. After the initiation of the control of the entire system takes place, the condensing unit control system takes over and supervises the operation of the equipment. This control system has all the necessary safety devices for the major load and contains a timer with a relay to prevent rapid short cycling. The timer unit allows for a 5-minute delay between the cycles of the condensing unit.

It should be noted that many control systems will operate completely on low voltage and hence be very similar to the simple residential control systems.

Control Systems for Air-Cooled and Water-Cooled Packaged Units

Many air-cooled and water-cooled packaged units are designed strictly for commercial and industrial use. These packaged units are almost always used

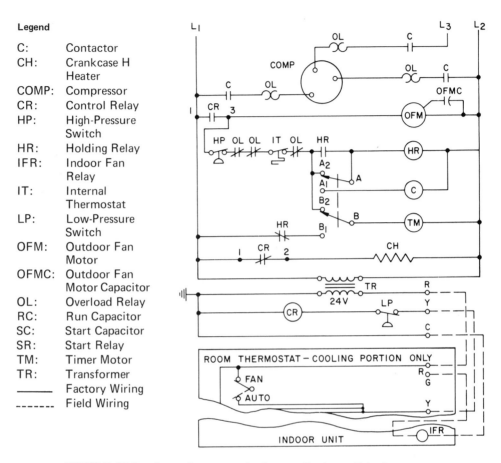

Legend

C: Contactor
CH: Crankcase H
         Heater
COMP: Compressor
CR: Control Relay
HP: High-Pressure
         Switch
HR: Holding Relay
IFR: Indoor Fan
         Relay
IT: Internal
         Thermostat
LP: Low-Pressure
         Switch
OFM: Outdoor Fan
         Motor
OFMC: Outdoor Fan
         Motor Capacitor
OL: Overload Relay
RC: Run Capacitor
SC: Start Capacitor
SR: Start Relay
TM: Timer Motor
TR: Transformer
———   Factory Wiring
- - - - Field Wiring

FIGURE 12.8.   Control system of a 5-ton split air-conditioning system with a fan coil unit. Reproduced by permission of Carrier Corporation. © 1977 Carrier Corporation.

in a free-standing installation put directly in the area that is to be conditioned, without any duct work.

This type of unit is equipped with a line voltage control system that is totally housed in the equipment.

The unit has a manual switch that allows it to be operated by this switch and a common line voltage thermostat. The three-position switch has an "off" position, a "fan only" position, and a "cool" position. The "cool" position of the switch is connected in series with a line voltage thermostat. It cuts the compressor on and off as needed by energizing a contactor or magnetic starter. If the unit is large, the evaporator fan motor is also controlled by a magnetic starter. The smaller evaporator fan motors, being single phase, are controlled by the switch.

All the safety devices are connected in series with the coil of the contactor. In most cases crankcase heaters are used on this type of system and are energized on the "off" cycle of the compressor. Figure 12.9 shows a pictorial diagram of a common control system used on this type of equipment.

Pump-down Control Systems

A large commercial and industrial air-conditioning condensing unit can be controlled by various types of initiation. The most common type of control is a pump-down system that incorporates a solenoid valve in the liquid line of the equipment. The valve opens and closes on the direction of the overall control system of the structure. The action of the solenoid valve determines if the condensing unit is to operate. If the solenoid valve is open, the unit will operate; if the solenoid closes, the unit will continue to operate until the low-pressure switch cuts the compressor off. Figure 12.10 shows a simple pump-down hookup of a condensing unit. The figure shows the solenoid valve being controlled by the thermostat and the action of the condensing unit controlled by the low-pressure switch. Some condensing units use a solenoid relay mounted in the unit to control the solenoid valves.

The pump-down control method is in common use on commercial and industrial systems because it will not allow refrigerant to feed to the coil on the "off" cycle and migrate back to the compressor. The solenoid valve in this type of system can be controlled by almost any type of temperature-sensing device that can be mounted in the desired portion of the system. Or the valve can be controlled by the equipment to prevent the valve from operating the equipment in unsafe conditions.

12.5 TOTAL COMMERCIAL AND INDUSTRIAL CONTROL SYSTEMS

The commercial and industrial control system is complex and difficult to understand because it controls virtually the total structure. It has the responsibility of controlling the zones of the structure, the air-conditioning equipment, the heating equipment, the evaporator fan motor, and the temperature of the entire structure. These control systems are usually designed and installed by one of the major control manufacturers.

The controls of a large system can be electric, pneumatic, or electronic. The electric control system is not used as often as the pneumatic and elec-

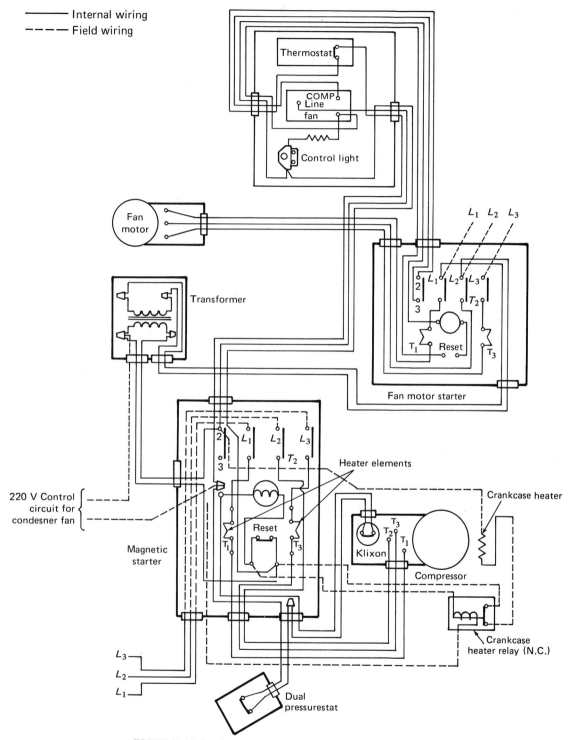

FIGURE 12.9. *Pictorial diagram of the control system of an air-cooled packaged unit with a remote condenser. Reproduced by permission of Carrier Corporation. © 1977 Carrier Corporation.*

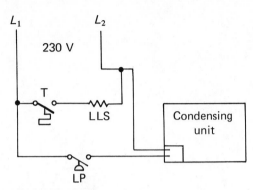

FIGURE 12.10. *Simple control system for the pump-down cycle of a condensing unit.*

Legend

T: Thermostat
LLS: Liquid Line Solenoid
LP: Low-Pressure Switch

tronic because the devices must be larger and do not maintain as accurate control.

The pneumatic control system is operated by air pressure, which opens and closes switches and valves by air pressure. An example of a pneumatic control system is shown in figure 12.11.

The electronic control system is shown in figure 12.12. It controls the temperature in the zones by electronic devices.

These diagrams were shown here just as examples of some of the control systems in use today. Their circuitry will not be discussed.

12.6 TROUBLESHOOTING CONTROL SYSTEMS

The service mechanic must develop a systematic method for finding the faults in a control system or in any electric part of the equipment. The responsibility of finding and correcting the problem in a control system falls on the person attempting to repair the equipment. This person is usually the service mechanic. It is no easy task to diagnose problems in electric systems, especially the larger commercial and industrial equipment. The smaller residential and commercial equipment, fortunately, is built and designed with a simpler control system and fewer components than the larger equipment.

Service personnel must understand how, when, and why a certain piece of equipment operates as it does. The service mechanic has many important tools that can be used to find and correct the problem in the equipment. The

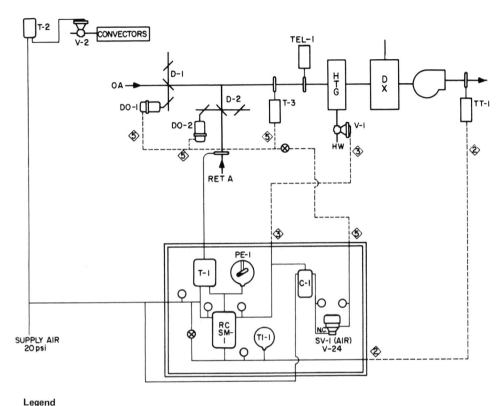

Legend

Ret A:	Return Air	T-2:	Room Thermostat
OA:	Outdoor Air	T1-1:	Discharge Air Thermometer
C-1:	Minimum Position Controller	V-1:	Hot Water Valve
DO-1	Outdoor Air Damper Controller	V-2:	Room Hot Water Valve (Convector)
DO-2:	Return Air Damper Controller		
D-1:	Outdoor Air Damper	TT-1:	Discharge Air Thermometer
D-2:	Return Air Damper	SV-1:	Air Solenoid Valve (Outside Air)
PE-1:	Cooling Controller		
T-3:	Mixed Air Thermostat	TEL-1:	Freeze Protector
RSCM-1:	Receiver Controller	-------	Air Lines Field Installed
T-1:	Return Air Thermostat	——	Supply Air Lines

FIGURE 12.11. Pneumatic control system. Diagram from Johnson Controls, Inc.

wiring diagram of the equipment is extremely important to the service mechanic because it tells how the equipment operates. The diagram also tells
the service mechanic the components in the electric system and their location in the system. The schematic diagram is usually the easiest tool to use
for diagnosing problems. However, in many cases manufacturers will give a
pictorial diagram or a combination of a pictorial and schematic, with the

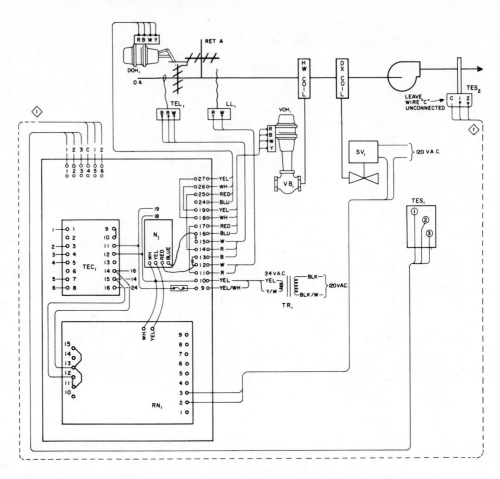

Legend

DOH:	Outdoor Air Damper	TR:	Transformer
TEL:	Outdoor Air Thermostat	TEC:	Main Controller
LL:	Low Limit Thermostat	N:	Outdoor Air Adjustment Relay
VB:	Hot Water Valve	RN:	Staging Relay
VOH:	Valve Operator	TES1:	Room Sensor
SV:	Cooling Solenoid Valve	TES2:	Discharge Air Sensor

FIGURE 12.12. Electronic control system. Diagram from Johnson Controls, Inc.

equipment. No matter what type of diagram is furnished with the equipment, the service mechanic must find and repair the malfunction in the equipment. The voltmeter, ammeter, and ohmmeter are all important tools in accomplishing this task.

It is essential that service personnel be able to read wiring diagrams and use electric meters to diagnose electrical problems and make the necessary repairs. Electric components are usually the cause of any electrical problems in the system. Therefore, the service mechanic must be efficient at correctly diagnosing the condition of the electric components of the system, such as the compressor, electric motor, relays, contactors, transformers, thermostats, and pressure switches. Of course, the mechanic must at all times be careful not to overlook the small things that homeowners or inexperienced service mechanics could have done to an electric system, such as changing wires, jumping controls out, and removing components they felt were unnecessary.

Diagnosis of Electric Components

The diagnosis of electric components is a fairly easy task for the mechanic who understands the particular device (see the earlier chapters in this book). In many cases service mechanics assume the jobs of parts changers, instead of actually diagnosing each component that has a possibility of being faulty. A good service mechanic will change some electric components that are perfectly good, but this should happen only rarely.

Electrical problems are very common throughout the industry and are said to cause 80% to 90% of the reported system failures. Thus it is important for service personnel to be able to diagnose and correct electrical problems in a system.

Motors.   The electric motor is the largest load used in most heating, cooling, and refrigeration systems. The compressor motor is the most expensive component of the system and care should be taken to correctly diagnose this component because of the expense that can be needlessly incurred. In fact, all electric motors are expensive and should be accurately diagnosed. For details about electric motors, refer back to chapters 6 and 7.

Switches.   Electric switching devices can cause many of the electrical problems encountered by service mechanics. These devices are used to control some load in the system. The relay and contactor are basically the same device except for their size and ampacity. These devices must be checked for two possible malfunctions. First, the contacts or mechanical linkage can be bad. Second, the coil that closes the contacts can be faulty. These devices are easily diagnosed for problems if the mechanic understands their operation and construction.

Thermostats and pressure switches are switches that are controlled by the temperature of the medium surrounding them or the pressure at the point they are connected to the system. A simple thermostat or pressure switch can easily be diagnosed because they have only one set of contacts. But the mechanic must be absolutely certain that the problem is not due to some other device that is malfunctioning.

The low-voltage thermostat used on the smaller systems is more difficult to diagnose because of its complexity and the fact that it is used for several purposes on a heating and cooling system. The service mechanic should be careful not to condemn as faulty a thermostat that is merely out of calibration by a few degrees.

There are many other electric components used in control systems that have not been covered in this review. However, they should be easily diagnosed when the mechanic studies the component along with the wiring diagram.

Diagnosis of Control Systems

In many cases the components are at fault when the system is not operating properly. But it is sometimes difficult to identify the faulty component unless the mechanic uses the wiring diagram to identify the circuit that is not operating properly and then goes to the load or to the device that controls the load. A mechanic can usually identify the load that is not operating properly. Therefore, the circuit should be checked for any malfunctions or faulty components.

Residential Systems. Figure 12.13 shows a basic circuit in a residential air conditioner. The circuit contains the contactor and the compressor, with the contactor controlling the load. If the compressor is not operating, it is logical to assume that the problem is with the power supply to the contactor or the

FIGURE 12.13. Basic circuits in a residential air-conditioning system.

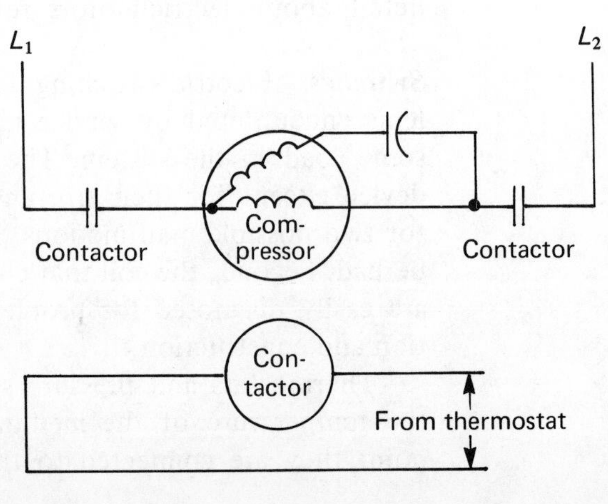

Legend

C: Contactor
COMP: Compressor
CRC: Compressor Running Capacitor
CFM: Condenser Fan Motor
CFMC: Condenser Fan Motor Capacitor
IFR: Indoor Fan Relay
IFM: Indoor Fan Motor
HP: High-Pressure Switch
LP: Low-Pressure Switch
TR: Transformer

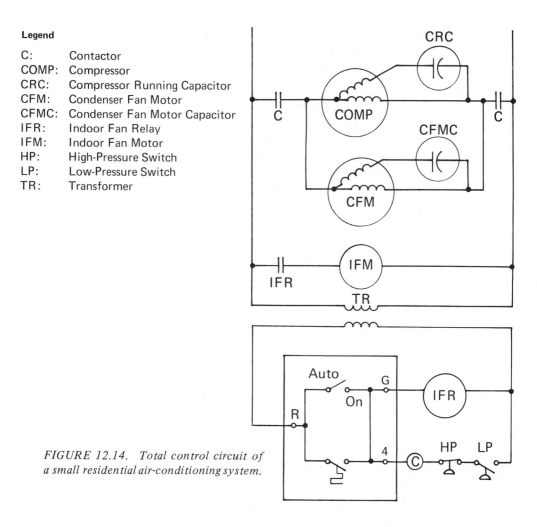

FIGURE 12.14.   Total control circuit of a small residential air-conditioning system.

unit. The first check should be the incoming power to the contactor, usually to a terminal board. If it is available, continue the voltage check to the compressor. If voltage is not available to the compressor, check the contacts and make certain of their position. If the contactor is closed and the compressor is being supplied with the correct power, diagnose the compressor, making certain to consider the overload if it is internal. Check the contacts of the contactor if the contactor is closed but no voltage is going through its contacts. However, if the contacts are not closed, you will have to look at the circuit that controls the contactor coil.

Total Circuit. The total control circuits of a residential air-conditioning system are shown in figure 12.14, including the total wiring diagram. The methods used in the previous paragraphs can be followed here. The low-voltage control circuit can be checked out if the contactor is not closing, assuming the compressor is good and the contactor is not closed. The first check that should be made is the low-voltage power supply to ensure that voltage is available to the control circuit. The line voltage power supply has been checked and is correct.

Once it is determined that voltage is available to the control circuit, the only items left to check are the devices in the circuit. In the control circuit there are four devices that should be checked: the low-pressure and the high-pressure switch, the thermostat, and the contactor coil. The contactor coil can be checked by reading the voltage available to it. If 24 volts are available but the contactor is not closed, then the coil or mechanical linkage of the contactor is bad and should be replaced. The other three components should be checked for an open circuit. This can be done easily with an ohmmeter or voltmeter. Do not overlook any problem that could have the high-pressure or low-pressure switches opened (low refrigerant charge, dirty condenser, or a bad condenser fan motor).

Hopscotching.   One frequently used procedure for troubleshooting an electric circuit is called **hopscotching**. Figure 12.15 shows an electric circuit that we will use as an example to follow the procedure. The contactor coil in the circuit closes, starting a compressor. Let us assume that we are getting voltage to the system but not to the contactor coil. If the correct voltage is not available to the coil, one of the switches in the contactor coil circuit would have to be open. The question is which one. First connect one lead of a voltmeter to L_2. The other lead of the voltmeter should be placed at point A in figure 12.15. If line voltage is read, the switch at A is closed and not at fault. Then hop to points B, C, and D with the meter leads. At the point where voltage is not read, that switch is open.

This procedure of troubleshooting is only one of many used in the in-

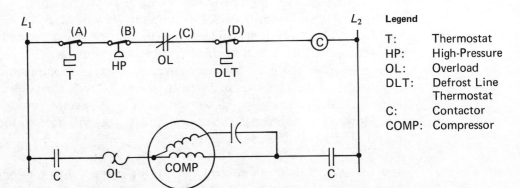

FIGURE 12.15.   *Simple electric circuit for hopscotching.*

dustry. Good service technicians will usually develop their own procedures as they gain experience with the various types of equipment.

Industrial Systems. Commercial and industrial control systems have many more components and are more complex than the simple residential control systems. Therefore, it is more difficult to diagnose and correct the problems in these circuits. First check the power supply to the equipment and the power supply to the control circuit. If there is power, use the wiring diagram and follow a systematic method for troubleshooting the system. The complexity and number of components should not cause you any concern. Each circuit of the diagram can be and should be treated individually.

Service personnel soon learn their own methods of troubleshooting equipment which allows for faster and more efficient service calls. The one item that takes its toll on service mechanics is a callback. A callback is a return trip to a piece of equipment that a mechanic has not fixed properly. This hurts the efficiency of the mechanic as well as the employer because of the added expense, which cannot be charged to the consumer. Service mechanics should try to eliminate all possibility of callbacks. If a mechanic does the right kind of job on the first call, there is no callback.

One procedure for troubleshooting any air-conditioning, heating, or refrigeration system is the following:

1. Check the power supply to the equipment.
2. If no power is available, correct the problem.
3. Make a thorough check of the system to determine which load or loads are not working properly.
4. Locate the loads on the wiring diagram that are not operating properly and begin checking these circuits first.
5. Each device in that circuit is a possible cause of the problem and should be checked until the faulty device is found. (Do not disregard the load in the circuit.)
6. It may be necessary to move to other circuits to check the coils of some of the devices.
7. Never overlook the internal overload devices.

Most manufacturers make a troubleshooting chart available to air-conditioning dealers that handle their line of equipment. Figure 12.16 shows a troubleshooting chart of one manufacturer for a total cooling system. Other manufacturers develop troubleshooting charts on each different model of equipment they produce. If these charts are available, it is advisable that they be used, especially if you are just beginning your service career.

The heating, cooling, and refrigeration industry depends as much on the service aspects as on the selling of new equipment. Service work is a necessity if a company is going to be successful. No consumer wants an air-conditioning

TROUBLE-TRACER*
Summer Air Conditioning and Heat Pump

TAPPAN
AIR CONDITIONING DIVISION

Satisfactory operation of an adequately sized air conditioning system depends on proper discharge pressure, suction pressure, current draw and air distribution. The service man must be able to determine these in order to properly diagnose system operation

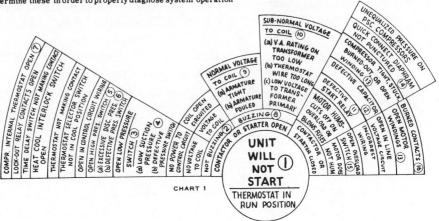

CHART 1

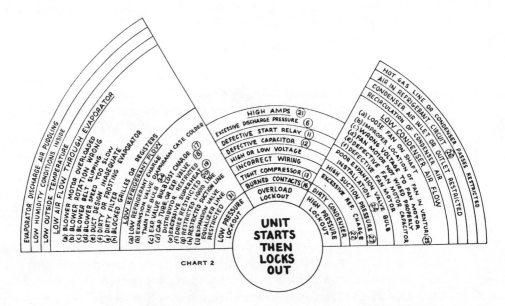

CHART 2

Center of chart states basic trouble. Begin there.

Remaining partial circles state various contributing troubles to specific trouble. Determine contributing trouble or troubles before attempting correction.

Adjacent partial circles state specific troubles that could cause basic trouble. Determine which of these specific troubles exist before proceeding further.

Numbers in chart are for explanation and guidance. Where no numbers are shown, the statement is considered to be self-explanatory.

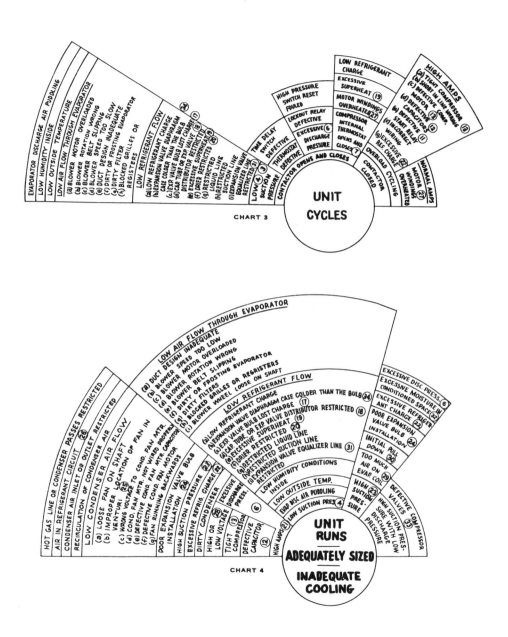

FIGURE 12.16.   Manufacturer's troubleshooting chart. Chart courtesy of Tappan Air Conditioning Division.

and heating system if he or she cannot get the proper service and maintenance. Remember, too, that service personnel always reflect the employer
when they are sent out on the job. They should always try to present themselves in a manner that will sell them as well as their employer. The appearance and attitude of a service mechanic are just as important as the ability to
diagnose and correct the problem that the homeowner is having.

SUMMARY

There are certain similarities among control systems used on the smaller
heating and air-conditioning equipment. The contactor in almost all cases
will control the compressor and the condenser fan motor. In some cases the
condenser fan motors are controlled by a thermostat to maintain a constant
head pressure of the system, but usually a contactor controls a two-speed
fan motor. The evaporator fan motors are usually controlled by some temperature on the heating cycle and an indoor fan relay on the cooling cycle.
The safety components are in series with the contactor controlling the compressor because it is the largest, most important, and most expensive load in
the heating and air-conditioning system. From a familiarity with the common basic control circuits, an installation or service mechanic can easily
adapt to the different control systems used in the industry. There is almost
always some common type of control used in all control circuits.

The advanced control system required on the larger commercial and industrial equipment is more complex and sophisticated than the smaller
equipment controls because of its size and operation. In many large structures the control system controls the structure and only maintains a certain
temperature for the heating and cooling medium. Therefore, the main control system signals for the cooling or heating equipment to operate. Once
the equipment has been started, the control system of the equipment supervises the operation of the unit, including the capacity control.

There are basically two methods used to start condensing units and air-
conditioning equipment. The pump-down method uses a solenoid valve to
start and stop the flow of refrigerant to the evaporator. The other method
used to start the equipment is by supplying a certain voltage to the equipment when it needs to be started. Large control systems have many basic
circuits that are common to equipment throughout the industry.

The servicing of any air-conditioning or heating system is important because sooner or later the equipment will malfunction and need attention.
Service mechanics must develop a systematic approach to diagnosing the

faulty equipment and making the necessary repairs. A good beginning for any service mechanic is to be able to understand and diagnose all the components of the system. Service mechanics should use the many tools that are available to them when servicing equipment, such as wiring diagrams, testing equipment, and any service manuals that are available. No matter how large or how small an air-conditioning, heating, or refrigeration system is, the mechanic can break down the control system and the electric system into basic circuits and then check the circuits on an individual basis. This is much easier than attempting to treat the total diagram as one large circuit. By using a systematic approach to troubleshooting electric systems, service mechanics can accomplish any troubleshooting task within reason.

QUESTIONS

1. Explain briefly the operation of the unit in figure 12.17.

2. Explain briefly the operation of the unit in figure 12.18.

3. Outline one procedure for troubleshooting a control system.

4. Name three components that would be common to all control systems.

5. True or false. All evaporator fan motor circuits are alike for the smaller residential systems.

6. On residential systems the condenser fan motor is controlled by _____.

7. True or false. Safety controls used in the smaller systems are connected in parallel with the contactor coil.

8. Why is capacity control used on the larger systems?

9. The part winding motor is used to _____.

10. True or false. In the larger systems fan motors are generally designed to operate at all times.

11. A constant head pressure is maintained in a system by the _____.

12. In the larger systems a low-voltage control system would be used to _____.

13. What is the purpose of the main control system in the larger industrial and commercial units?

14. What is a pump-down control system? How does it operate?

15. The controls of a large system can be _____, _____, or _____.

16. True or false. The compressor is the most expensive component of an air-conditioning system.

17. What is the first check that should be made in diagnosing a basic control system?

18. Describe what is meant by the term "hop-scotching."

19. Outline one procedure you might use for troubleshooting an air-conditioning, heating, or refrigeration system.

20. Use figure 12.17 to answer the following questions:
    a. What is the problem with the unit if the contactor is closed but the compressor is not operating?
    b. What safety devices could prevent the contactor from closing if the thermostat, selector switch, and the two overloads are closed?

21. Use figure 12.18 to answer the following questions:

a. What type of thermostat is used in the diagram?
b. What are the possible problems with the control system if the control relay is energized but the compressor is not operating?
c. What is the problem with the unit if it will not come out of the defrost cycle?

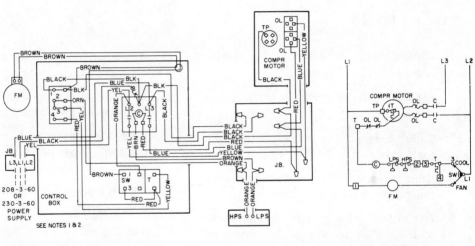

Legend

C:	Compressor Contactor	☐ :	Terminal Board Connection
COMPR:	Compressor	———	Control Wires
F:	Fan Contactor	━━━	Power (factory) Wires
FM:	Fan Motor	------	Field Wires
HPS:	High-Pressure Switch	— — —	Alternate CSR Wiring
IT:	Internal Thermostat		
LPS:	Low-Pressure Switch	**Notes**	
OL:	Overload Protector		
SW:	Selector Switch	1.	Fan motors furnished with inherent thermal protection; compressor motors thermal and overload protection.
T:	Thermostat		
⬡ :	Component Connection, Marked		
◇ :	Component Connection, Unmarked	2.	Control terminal block for connection of Class 1 control circuit.

FIGURE 12.17. Diagram for questions 1 and 21. Reproduced by permission of Carrier Corporation. © 1977 Carrier Corporation.

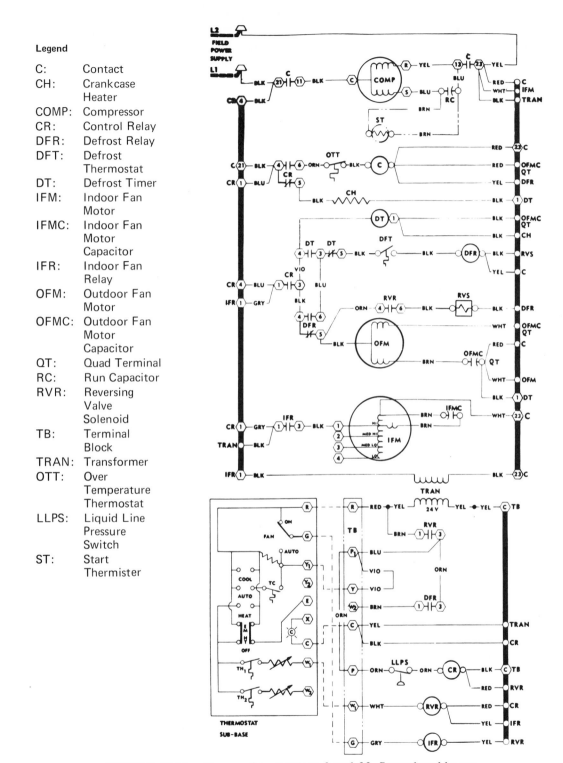

Legend

C: Contact
CH: Crankcase Heater
COMP: Compressor
CR: Control Relay
DFR: Defrost Relay
DFT: Defrost Thermostat
DT: Defrost Timer
IFM: Indoor Fan Motor
IFMC: Indoor Fan Motor Capacitor
IFR: Indoor Fan Relay
OFM: Outdoor Fan Motor
OFMC: Outdoor Fan Motor Capacitor
QT: Quad Terminal
RC: Run Capacitor
RVR: Reversing Valve Solenoid
TB: Terminal Block
TRAN: Transformer
OTT: Over Temperature Thermostat
LLPS: Liquid Line Pressure Switch
ST: Start Thermister

FIGURE 12.18. Diagram for questions 2 and 22. Reproduced by permission of Carrier Corporation. © 1977 Carrier Corporation.

Glossary

air-cooled packaged unit: A unit that is made in one complete unit with an air-cooled condenser mounted with the unit or remotely.

alternating current: Electron flow that flows in one direction and then reverses at regular intervals; produced by cutting a magnetic field with a conductor. The most common type of power supply used in the heating, cooling, and refrigeration industry.

ambient air temperature: The temperature of the air surrounding any device.

ammeter: An electric meter used to measure the amperes that are present in a circuit. Ammeters are made as an in-line ammeter or clamp-on meter.

ampere: The amount of current required to flow through a resistance of one ohm with a pressure of one volt, measured with an ammeter.

anticipator: A component of a thermostat that anticipates the temperature of an area and will stop or start the cooling or heating equipment to prevent the thermostat from overshooting the desired temperature.

apparent wattage: The power that is calculated by multiplying the voltage times the amperage of a circuit.

armature: The portion of a contactor that moves; connected to a set of contacts that causes a completed circuit when the armature is pulled into the magnetic field produced by the coil.

atom: The smallest particle of an element that can exist and maintain any identification; can combine with other atoms to form new substances.

back electromotive force: The amount of voltage produced across the starting and common terminals or connections of a single-phase motor.

ball bearing: An antifriction device that is used to allow free turning and support of the rotating member of the device. It consists of an outer ring and inner ring with races and has steel balls sandwiched between the rings.

bearing: The part of a rotating machine or motor that allows free turning of its rotating parts with little friction.

bimetal overload: A simple thermal overload that breaks the power supplying a small motor directly. When it reaches a high enough temperature, the bimetal opens by warping; when the overload cools, the bimetal will warp back and close the circuit.

bimetal relay: An overload device that opens a set of contacts by a temperature that corresponds to current draw of a load.

breaker: A device usually used in a breaker panel that is capable of being used as a disconnect switch and an overload for the circuit it is supplying power to.

breaker panel: An electric panel that houses breakers used to distribute power to circuits in the structure.

callback: A return trip to a piece of equipment that was not properly serviced the first trip.

capacitive reactance: A resistance caused by using capacitors with motors; when a capacitor is put in a circuit, it resists the voltage change, causing the amperage to lead the voltage.

capacitor: A device that consists of two aluminum plates with an insulator between the plates; used to boost starting torque of single-phase motors. The two types of capacitors are electrolytic or starting and running or oil-filled.

capacitor-start–capacitor-run motor: A high starting torque and good running efficiency motor; it uses a starting capacitor to increase starting torque and a running capacitor to increase running efficiency.

capacitor-start motor: A motor that uses a starting capacitor to increase the starting torque.

charge: A condition in which an imbalance of protons and electrons exists in an atom.

closed circuit: A complete path for electrons to follow.

compressor: A device that is used in a mechanical refrigeration system to compress the refrigerant. Most compressors are rotated by an electric motor. The motor may be external or be an integral part of the compressor.

condensing unit: A portion of a split air-conditioning or refrigeration system that is mounted outside and contains the compressor, condenser, condenser fan motor, and controls for these components; most used today are air-cooled. It takes the cool suction gas from the evaporator, compresses it, condenses it to a liquid, and forces it back to the evaporator.

conductor: A wire that is used for the path of electric flow. Most electric conductors are copper or aluminum.

contactor: A device that opens and closes a set of electric contacts by the action of a solenoid coil; composed of a solenoid and the contacts.

continuity: A complete path for electrons to follow in a circuit or component.

control circuit: A circuit that controls some load in the entire control system, whether it be a relay or contactor coil or a major load.

current: Electrons flowing in an electric circuit, measured in amperes.

current electricity: Electricity that results from the electron being displaced and moving back to the atom.

current overload: An overload that opens a set of contacts on high current draw and allows them to close when the current draw has decreased. It usually is a pilot duty device.

current relay: A relay that is opened or closed by the starting current of an electric motor. The relay allows a starting capacitor and starting winding to drop out or drop in the starting circuit.

cut-in pressure: The pressure at which a pressure control will close, starting the device it is controlling.

cut-out pressure: The pressure at which a pressure control will open, stopping the device it is controlling.

cycle: One complete cycle of alternating current is the production of a positive and negative peak.

deenergize: To stop the electron flow to an electric device.

delta transformer: A three-phase transformer that has the ends of each of its windings connected together to form a triangle. The delta transformer produces a high leg on one of its power legs.

delta winding: A winding layout of some three-phase motors, where the beginning of the windings is connected to the ending of the windings.

dielectric: The substance that is between the plates and fills the case of a capacitor. The substance is a nonconductor of electricity.

differential: The difference between the cut-in and cut-out point of a control. This can be applied to thermostats, pressure switches, and most controls.

direct current: An electron flow in only one direction; used in the industry only for special applications such as solid-state modules, electronic air filters, and special applications.

direct drive: A method of transferring rotating motion of a motor to a device that must be turned. This type of hookup connects the motor directly to the device that must be turned and rotates the same revolutions per minute as the motor.

disconnect switch: A switch that is used to disconnect the power supply to a piece of equipment; it is sometimes referred to as a safety switch.

distribution center: An electric panel used to distribute electric supply to several places in a large structure; can be of fusible or circuit breaker design.

effective voltage: Alternating current, with its many reversals and peaks, never peaks at a constant value. The effective voltage is the working voltage of alternating current. The effective voltage is 0.707 times the peak voltage.

electric circuit: A path for electrons to follow; the circuit may be open or closed depending on the position of its switches.

electric energy: Energy that is produced by a movement of electrons. The energy can be produced by chemical, light, thermal, or mechanical means.

electricity: Energy that is capable of producing an electron flow. An unbalanced condition that results when an electron can be easily displaced from an atom.

electric meter: A device used to measure some electrical characteristics of a circuit such as the voltage, amperage, resistance, or wattage.

electric power: The rate at which electricity is being used, measured in watts.

electric pressure: Another term used to refer to electromotive force, potential difference, and voltage.

electric switch: A device that opens or closes to control some load in an electric circuit; it can be opened or closed by temperature, pressure, humidity, flow, and manual means.

electrodes: The dissimilar metal conductors in a battery that produce a small difference of potential.

electrolyte: The chemical paste between the electrodes of a battery and some capacitors.

electromagnet: A magnet produced by coiling wire around a metal core.

electromotive force (emf): The difference of potential that forces electrons through a resistance. *CH I QUEST 7*

electron: Particles that orbit around the nucleus of an atom and have a negative charge.

element: A substance that has weight, takes up space, and cannot be broken down by chemical means.

energize: To apply voltage to an electric device.

energy efficiency rate (EER): The means of *P. 13* measuring an air conditioner for its efficiency by stating how many Btu's of cooling are available from one watt.

factory-installed wiring: The wiring installed in a piece of equipment at the factory; usually the connections between the components in the control panel and the system components in the unit itself.

factual diagram: A wiring diagram that is a combination of the pictorial and schematic diagrams.

field of force: The area around a magnet that is affected by the strength of the magnet.

field wiring: The wiring that must be installed in the field by the installation mechanic.

flux: The magnetic lines of force of a magnet *P. 120* that connect the north and south poles of the magnet.

free electron: Electrons that are easily removed from the outer orbits of atoms.

frequency: The number of complete cycles per second of alternating current.

fuse: A device that breaks a circuit when its

ampere rating is exceeded; constructed of two ends or conductors with a piece of wire that will melt and break the circuit on an overload.

fusible load center: An electric panel that supplies circuits with power and protects them with fuses.

gas pack: A unit that heats in the winter by using gas as its fuel and cools in the summer by using electric power.

head pressure: The discharge pressure of a refrigeration system, sometimes called high-side pressure.

heater: An electric load that converts electric energy to heat.

heat pump: A refrigeration system that reverses the flow of refrigerant in the normal refrigeration cycle, which allows the unit to cool in the summer and heat in the winter.

hermetic compressor motor: A motor that is designed for single- and three-phase operation and is totally enclosed in a shell with refrigerant and oil.

hertz: The number of complete cycles per second of alternating current; more widely accepted than the term "frequency."

hopscotching: A troubleshooting procedure for electric circuits that is accomplished by jumping from one component to another.

hot-wire relay: A relay that is opened or closed by a thermal element that senses the starting current of the motor. The relay allows a starting capacitor or starting winding to drop out or drop in the starting circuit. This type of relay also has a built-in means of overload protection.

humidistat: A device that is used to control humidity; it uses a moisture-sensitive element to control a mechanical linkage that opens and closes an electric switch.

impedance: The sum of the resistance and reactance in an alternating current circuit.

indoor fan relay: An electric relay that starts and stops an indoor fan on cooling, electric-heating, and heat pump systems.

induced magnetism: The magnetism induced around a current-carrying conductor. *p. 121*

inductance: A property of an alternating current circuit by which an electromotive force is produced in it by a variance in current.

inductive load: A load that starts with a larger ampere draw and reduces it as the load starts normal operation. The increase in the ampere draw initially is due to inductance.

inductive reactance: The opposition to the change in alternating current flow that produces an out-of-phase condition between voltage and amperage.

installation diagram: A diagram that shows little internal wiring but gives specific information as to terminals, wire sizes, color coding, and breaker or fuse sizes.

insulator: A material that retards the flow of electrons or electricity.

internal compressor overload: An overload that is embedded in the windings of a motor. Some internal overloads break the power to the motor directly while others merely open a set of contacts that is wired into an electric control circuit.

kilowatthour: The rate at which electric energy is being used at a specific time. Most electric utilities bill their customers in this method.

law of electric charges: Like charges repel and unlike charges attract.

line break overload: An overload that breaks the power going to the motor and is most commonly used on small motors.

line voltage: The voltage being supplied to the equipment as the power supply. FROM SOURCE

load: Electric devices that consume electricity to do useful work, such as motors, solenoids, heaters, and lights.

locked rotor amperes: The current a motor uses the instant it starts while the rotor is in a stationary position.

magnetic field: The area around a magnet in which the effect of magnet can be felt.

magnetic overload: An overload device that senses the current draw of a load by the magnetic field produced, which is proportional to the current draw. The device will open a set of contacts on high current draws and allow them to close when the ampere draw returns to normal.

magnetic starter: A device that opens and closes its contacts when a solenoid is energized. A means of overload protection is provided. It is the same as a contactor except for the overload protection.

magnetism: The ability of two pieces of iron to be attracted to each other by physical means or electrical means.

matter: The substance of which all physical objects consist.

measurable resistance: The actual resistance of a circuit or component measured with an ohmmeter.

microfarad: The unit of measurement used to measure the strength of a capacitor.

molecule: The smallest particle into which an object can be divided and maintain the properties of that object.

molten-alloy relay: An overload device that opens a set of contacts by thermal energy. This type of device allows the temperature produced by the starting current of a load to be transferred to a molten-alloy device. When it reaches a certain temperature, it will melt the solder around the device, causing it to slip and open the contacts; when it cools it will harden again and the relay must be manually reset.

motor: A device used to create a rotating motion and drive components that require rotating motion. Electric energy is changed to mechanical energy by magnetism, which causes the motor to turn. p 46

negative charge: The result of electrons joining atoms.

neutron: The neutral particle in the nucleus of an atom.

noninductive load: A load that has only resistive qualities with no inductive qualities. An electric heater and incandescent lighting are two common types of noninductive loads.

normally closed: The position of a set of contacts or other electric devices that are closed when the device is deenergized.

normally open: The position of a set of contacts or other electric devices that are open when the device is deenergized.

nucleus: The central part of an atom composed of protons and neutrons.

ohm: The amount of resistance that will allow one ampere to flow with a pressure of one volt.

ohmmeter: An electrical meter used to measure the resistance of a circuit or electric component.

Ohm's law: The relationship between current, electromotive force, and resistance in an electric circuit.

open circuit: A circuit without a complete path for electrons to follow.

out of phase: The voltage and current are not working together.

overload: A device that is used to detect a high ampere draw of some electric load and break the controlling circuit, stopping the load.

overshoot: The additional heating or cooling that has been delivered to the conditioned space after the thermostat contacts have opened.

packaged air-conditioning unit: A system built with all components in one unit except for the field wiring.

parallel circuit: An electric circuit that has more than one path for current flow.

peak voltage: When the voltage reaches its peak in an alternating current circuit.

permanent magnet: A piece of material that is magnetic by physical means. Iron, nickel, cobalt, and chromium are materials that can easily be magnetized and will maintain their magnetism for a period of time.

permanent split-capacitor motor: An electric motor, widely used in the industry, that has a moderate starting torque and good running efficiency.

phase: The number of currents alternating at different times in an alternating current circuit.

pictorial diagram: A wiring diagram that shows the actual internal wiring of a unit, much like a picture taken of a control panel. It is also called a line or label diagram.

pilot duty: A term used to refer to an electric device that indirectly controls a major load because of its large ampere draw but controls it directly through a device that is capable of carrying the load.

pole: One set of electric contacts either in an automatic device or a manual switch. Electric devices such as relays, contactors, switches, and breakers can be purchased with one or many poles.

positive charge: The result of electrons leaving an atom.

potential coil: A coil energized by a voltage being applied to it. It can be designed to operate on 24, 110, 208/230, or 460 volts. These coils are used on relays, contactors, and magnetic starters.

potential difference: Two points that have a difference in electric charge; the electric difference between two points in an electric circuit.

potential relay: A relay that uses the back electromotive force of a motor to drop out the starting apparatus when the motor reaches 75% of full speed.

power factor: The ratio of true power to apparent power, usually expressed as a percentage.

pressure switch: A device that opens or closes a set of contacts when a certain pressure is applied to the diaphragm of the switch.

proton: A positively charged particle in the nucleus of an atom.

pump-down control system: A control system that closes a solenoid to allow the compressor to pump all the refrigerant from the low side of the system into the high side. This system is used on large air-conditioning systems and some commercial refrigeration systems.

push-button switch: A switch that can be opened or closed by pressing buttons on the switch. Push-button switches come with a wide variety of purposes and labeling.

range: The operating ranges or limits of a control.

reactance: The resistance that alternating current encounters when it changes flow.

relay: A device that opens and closes a set of contacts when its coil is energized. The relay is much like the contactor except for its smaller size.

relief valve: A device that will open on a rise in pressure and release pressure to return a closed system to a safe operating condition and close when the pressure has decreased.

resistance: The opposition to the flow of electrons.

rotor: The rotating part of an electric motor.

running (oil-filled) capacitor: An electric device that is used to momentarily store electrons and create a second phase in the starting winding circuits of single-phase motors. This type of capacitor is designed to stay in the circuit whenever the motor is running as a means of heat dissipation.

safety device: Any device that is in a control system for the purpose of safer operation of a major load.

schematic diagram: A diagram that lays out the control system circuit by circuit and is composed of symbols representing components and lines representing their interconnecting wiring.

series circuit: An electric circuit that has only one path for electron flow.

series-parallel circuit: A combination of series and parallel circuits.

shaded-pole motor: An induction type of motor that does not incorporate an ordinary type of starting winding. It uses a band on one side to obtain a short-circuit effect that produces a rotating magnetic field. This motor has a low starting torque.

short circuit: An electric circuit that has no resistance.

short cycling: A term used to refer to a condition that occurs when a load is stopping and starting too frequently.

signal light: A light that is used to show when some electric component or circuit is energized by illuminating the light.

sine wave: A graphical representation of alternating current; a graph showing the sine function of all angles from 0 to 360 degrees.

sleeve bearing: An antifriction device that allows free turning and support of the rotating member of a device. It consists of a solid piece of bronze or babbit that is round and drilled to the diameter of the shaft. The bearing is sometimes called a plain bearing or bushing.

sliding armature: An armature that mounts between two slots in a contactor frame and moves up and down the slots when the contactor is energized.

snap action of a thermostat: The closing of a set of contacts of a thermostat with a snapping motion rather than with a light contact.

solenoid: A device that, when energized, will create a magnetic field and cause some action to an electric component. It opens and closes to control some element of a heating, cooling, and refrigeration control system.

solenoid valve: A valve that opens or closes by a solenoid coil being energized to pull a steel core into the magnetic field of the solenoid.

split-phase motor: An electric motor that has a running and starting winding. This is an induction type of motor.

squirrel cage rotor: The rotating part of an electric motor; its name is derived from its appearance of a squirrel cage. This type of rotor is used in split-phase, capacitor-start, shaded-pole, and three-phase motors.

staged system: A system that has more than one mode of heating or cooling operation.

staging thermostat: A thermostat that is designed to open and close more than one set of contacts to control several modes of heating or cooling operation.

starting capacitor: An electric device that is used to momentarily store electrons, creating a second phase in starting windings of single-phase motors. This type of capacitor is designed to stay in the circuit only a short period of time.

starting relay: A relay that is used to energize or deenergize the starting components of a single-phase motor.

star transformer: A three-phase transformer that has the ends of each winding connected to a common point. The star transformer produces a balance of all hot legs to ground.

star winding: A winding layout of some three-phase motors in which the end of the windings are connected together.

static electricity: Electricity that results from the electron being displaced and not returning to the original atom; usually results from friction.

stator: The stationary part of an electric motor.

swinging armature: An armature used in a contactor that is mounted on a line and moves up and down in a swinging motion.

switch: A device for making, breaking, or changing the connection in an electric circuit.

system lag: The difference in temperature between the point at which the thermostat closes and the point at which the thermostat starts to rise or fall.

thermal overload: An overload device that senses the current draw of a load by the heat produced, which is proportional to the current draw.

thermal relay: An overload device that determines current flow and opens a set of pilot duty contacts when an overload is indicated.

thermostat: A device that responds to a temperature change by opening or closing a set of electric contacts.

thermostat controlling element: The portion of a thermostat that reacts to temperature change by opening or closing the contacts through mechanical linkage. The two types of elements are bimetal and bulb.

three-phase motor: An induction type of motor that has a very high starting torque and requires no special starting apparatus. The motor must be operated on three-phase current.

throw: This refers to the number of positions of the movable contacts that will complete a circuit.

torque: The starting power of an electric motor.

transformer: A device that decreases or increases the incoming voltage to the desired voltage.

V-belt: The belt that connects the pulleys of a motor and the device that must be rotated and transfers the rotating motion from the motor to the device. V-belts can be purchased in several widths and almost any length.

volt: The amount of electric pressure required to force one ampere through a resistance of one ohm.

voltage: The difference in electric potential between two points.

voltage drop: The amount of voltage lost through any type of switching device or conductor.

voltmeter: An electric meter used to measure voltage.

water chillers: A refrigeration system that cools water that is pumped into other parts of the system to maintain the desired condition in a specific area.

water-cooled packaged unit: A unit that is made in one complete unit with a water-cooled condenser as an integral part.

watt: One ampere flowing with a pressure of one volt. The unit measurement of power.

wiring diagram: A systematic method of laying out the wiring that is interconnecting the control components within the control system; three types are schematic, pictorial or line, and installation.

Y transformer: A three-phase transformer that has a common junction point and forms a Y. This transformer hookup allows for a completely balanced load when using all hot legs and ground.

Index

Air conditioning, 44–80; loads, 45–59; motors, 46–47; solenoids, 48; heaters, 48–49; signal lights, 49; contactors, 50–53, 78; relays, 50–53, 78; magnetic starters, 53; switches, 53–58; safety devices, 58–60; transformers, 60; schematic diagrams, 60–74; pictorial diagrams, 74–76; installation diagrams, 76, 77, 79; control systems, 232–47; field wiring, 247–54; factory-installed wiring, 247; power wiring, 247–49; heating systems, 249–50; sizing wires and fuses, 250–51; control wiring, 251–54

Air conditioning, wiring diagrams: with series-parallel circuits, 30; components, symbols, and circuitry of, 44–80

Air-cooled packaged unit, def., 287

Alternating current, def., 287

Alternating current, 10, 17, 81–86; effective voltage, 83; cycles and frequency, 83; voltage-current systems, 83–84; phase, 84; inductance, 84; reactance, 84, 85; capacitive reactance, 85; inductive reactance, 85

Ambient air temperature, def., 287

American Wire Gauge (AWG), 99, 100

Ammeter, def., 287

Ammeter, 10, 34–37; clamp-on, 33, 35, 36–37; in-line, 35, 36

Ampere, def., 287

Ampere, 8, 10, 17

Anticipator, def., 287

Anticipators, 195–98; heating, 195–97; overshoot, 196; cooling, 197–98

Apparent wattage, 86, 287

Armature, def., 287

Atom, def., 287

Atomic theory, 1–3

Back electromotive force, def., 287

Ball bearing, def., 287

Ball bearings, 155, 156–157

Bearing, def., 287

Bearings, motor, 155–57
Bimetal overload, def., 287
Bimetal relay, def., 287
Bimetal thermostat, 188, 190–92
Breaker, def., 287
Breaker panel, def., 287
Breaker panels, 110, 112–16;
 construction, 112–14; installa-
 tion, 114–16
British thermal unit (Btu), 12

Callback, def., 288
Capacitive reactance, def., 288
Capacitive reactance, 85
Capacitor, def., 288
Capacitors, 133–35; starting, 133–
 34; microfarads, 134; running,
 134; electrolytic, 134; trouble
 shooting, 134–35
Capacitor-start-capacitor-run motor,
 def., 288
Capacitor-start-capacitor-run
 motors (CSR), 140–41; operation,
 140; troubleshooting, 140–41
Capacitor-start motor, def., 288
Capacitor-start motors, 138–40;
 open, 139; enclosed, 139–40
Charge, def., 288
Charges, positive and negative, 3–4
Chemical means, electricity
 through, 5–6
Circuit breakers: overload and,
 177; troubleshooting, 220–21
Circuits, 19–30; parallel, 19; series,
 19; basic concepts, 20–21; series-
 parallel, 20–30; compressor, 71;
 condenser fan motor, 71–74;
 safety controls, 74; timer relay,
 70–71
Coil, contactors, 171; troubleshooting, 219

Compressor, def., 288
Compressor circuits, 71
Compressor control circuits, 258
Compressor motor controls, 264
Compressors, 45–46, 47
Closed circuit, def., 288
Condenser fan motor circuits, 71–
 74
Condenser fan motor control
 circuits, 258–59
Condenser fan motor controls, 267
Condensing unit, def., 288
Condensing units, 233–39; simple
 control systems, air conditioners,
 235–36; complex control systems,
 air conditioners, 236–38; wiring,
 238; troubleshooting, 239
Conductor, def., 288
Conductors, 7, 8, 17
Components, electric motors, 149–
 65; starting relays, single-phase
 motors, 150; current or amperage
 relays, 150–52; potential relays,
 152–53; hot-wire relays, 154–55;
 motor bearings, 155–57; motor
 drives, 158–60; magnetic starters,
 160–63; push-button stations,
 163–64
Compound, 2–3
Contactor, def., 288
Contactors, 167–72; applications,
 167–68; operation, 168–69; coils,
 170, 171; contacts, 170–71, 172;
 troubleshooting, 171–72, 218–20,
 mechanical linkage, 172; repairing, 172
Contacts, contactors, 171–72;
 troubleshooting, 218–19
Continuity, def., 288
Control circuit, def., 288
Control devices, 186–216; troubleshooting, 217–31

Control systems, 166–85, 186; circuitry, 257–72; troubleshooting, 257, 272–85; compressor control circuits, 258; evaporator fan motor, 258–59, 266–67; condenser fan motor, 260, 267; safety, 261–62; residential units, 262–63; advanced, 263–70; compressor motor, 264; water chiller, 264; water-cooled condenser interlocks, 267; small commercial and industrial, 268; air-cooled packaged units, 268–70; water-cooled packaged units, 268–70; pump-down, 270; total commercial and industrial, 270–72; troubleshooting, 272–82; pneumatic, 273; electronic, 274; diagnosis, 276–82

Control systems, air conditioners, 232–47; condensing units, 233–39; packaged units, 239–47

Control wiring, air conditioners, 251–54; residential, 251; industrial, 252; sizing wire, 252–53

Cooling anticipators, 197–98

Current, def., 288

Current, 10, 17; flow, 10; direct, 10, 17; alternating, 10, 17; calculating for, parallel circuits, 26–29; calculating for, series circuits, 23–25

Current electricity, def., 288

Current overload, def., 288

Current relay, def., 288

Current relays, 150–52; operation, 150–51; troubleshooting, 151–52

Cut-in pressure, def., 288

Cut-out pressure, def., 288

Cycle, def., 288

Cycles and frequencies, 83

Deenergize, def., 288

Defrosting, time clocks and, 213

Delta transformer, def., 288

Delta winding, def., 288

Diagrams, air conditioners: installation, 76, 77; pictorial, 74–76; schematic, 60–74

Dielectric, def., 288

Differential, def., 288

Direct current, 10, 81, 289

Direct drive, def., 289

Disconnect switch, def., 289

Disconnect switch, 55, 56, 107–10; enclosures, 108; fusible and nonfusible, 108–10

Distribution center, def., 289

Distribution centers, 116–17

Effective voltage, def., 289

Effective voltage, air conditioners, 83

Electric charges, law of, 4

Electric circuit, def., 289

Electric energy, def., 289

Electricity, def., 289

Electricity: static, 4–5; through chemical means, 5–6; through magnetism, 6–7

Electric meter, def., 289

Electric potential, 8–9

Electric power, def., 289

Electric power: energy and, 12–13; calculating, 16

Electric pressure, def., 289

Electric pressure, 8, 9

Electric switch, def., 289

Electrodes, def., 289

Electrolyte, def., 289

Electrolytic capacitor, 134

Electromagnet, def., 289

Electromagnet, 121, 123

Electromotive force, def., 289

Electromotive force, 8, 9, 17
Electron, def., 289
Electronic control system, 274
Electrons, flow of, 4, 7, 17
Element, def., 289
Elements, 2
Energize, def., 289
Energy crisis, 13
Energy efficiency rating (EER), 13
    def., 289
Energy efficiency rating (ERR), 13
Evaporator fan motor, control
    circuits, 258–59, 266–67

Factory-installed wiring, def., 289
Factual diagram, def., 289
Field of force, def., 289
Field of force, 8
Field wiring, def., 289
Field wiring, 247–54
Flux, def., 289
Flux, magnets, 120, 121
Free electron, def., 289
Freon, pressure switches, 205, 207
Frequency, def., 290
Fuse, def., 290
Fuses, 59, 177; troubleshooting,
    220
Fusible load center, def., 290
Fusible load centers, 110–12

Gas pack, def., 290

Head pressure, def., 290
Heater, def., 290
Heaters, 48–49; resistance, 45, 46,
    50
Heating anticipators, 195–97
Heating, cooling, and refrigeration
    systems: installation, 98–118;
    sizing wire, 98–107; disconnect
    switches, 107–10; fusible load

centers, 110–12; breaker panels,
    112–17; magnetism and, 121–22;
    electric motors, 122; control
    systems, 166–85, 186
Heat pump, def., 290
Heat pump, thermostats, 202
Hermetic compressor motor, def.,
    290
Hermetic compressor motors, 144–
    46; operation, 144; terminals and
    troubleshooting, 144–46
Hertz, def., 290
Hertz, 83
High-pressure switches, 205
Hopscotching, def., 290
Hopscotching, 278–79
Horsepower (hp), 12
Hot-wire relay, def., 290
Hot-wire relays, 154–55; operation,
    154; troubleshooting, 154–55
Hot wires, 87
Humidistat, def., 290
Humidistats, 210

Impedance, def., 290
Impedance, 85
Indoor fan relay, def., 290
Indoor fan relay package, 251
Induced magnetism, def., 290
Inductance, def., 290
Inductance, 84
Inductive load, def., 290
Inductive reactance, def., 290
Inductive reactance, 85
Installation, 98–118
Installation diagram, def., 290
Installation diagrams, 76–77, 79
Insulator, def., 290
Insulators, 7–8, 17
Internal compressor overload, def.,
    290

Internal compressor overload, 180–81

Internal overload, troubleshooting, 223–25

Kiloampere, 10

Law of electric charges, def., 290

Law of electric charges, 4

Line break overload, def., 290

Line break overload, 178–79

Line voltage, def., 290

Line voltage overload, troubleshooting, 221–22

Line voltage, 25, 192, 194; troubleshooting, 225–26

Linkage, mechanical, contactors, 172

Load, def., 291

Loads, 45–49

Locked rotor amperes, def., 291

Low-pressure switches, 205–6

Low-voltage thermostats, 193–94, 195; troubleshooting, 226–29

Magnetic field, def., 291

Magnetic field, 120, 123

Magnetic overload, def., 291

Magnetic overload, 59–69, 182–83

Magnetic starter, def., 291

Magnetic starter, 53, 160–63; thermal relay, 162; molten-alloy relay, 162, 163; troubleshooting, 163

Magnetism, def., 291

Magnetism, 119–22; electricity through, 6–7; magnetic field, 120, 123; flux, 120, 121; induced, 121–22; permanent, 121; electromagnetic, 121, 123

Magnets, 119–22; horseshoe, 119–20

Manual switch, 54–55

Matter, def., 291

Matter, 1–2, 17

Measurable resistance, def., 291

Measurable resistance, 39

Mechanical linkage, contactor, troubleshooting, 219–20

Meters, 19–20, 31–41; differences among, 33; volt-ohm, 33, 37, 38; ammeter, 34–37; voltmeter, 37–38; ohmmeter, 38–41

Microfarad, def., 291

Microfarads, 134

Milliampere, 10

Molecule, def., 291

Molecule, 2, 17

Molten-alloy relay, def., 291

Motor, def., 291

Motor bearings, 155–57; ball, 155, 156–57; sleeve, 155, 157

Motor drives, 158–60; direct, 158–59; V-belt, 159–60

Motors, 45, 46–47; magnetism, 119–22; rotor, 123–24; strength, 125; types, 125–26; open and closed, 125–26; shaded-pole, 126–29; split-phase, 129–32; capacitors, 132–35; permanent split-capacitor, 135–38; capacitor-start, 138–40; capacitor-start-capacitor-run, 140–41; three-phase, 141–44; hermetic compressor, 144–46; components for, 149–65; troubleshooting, 230; diagnosis, 275

National Electrical Code (NEC), 99–101

Negative charge, def., 291

Neutron, def., 291

Noninductive load, def., 291
Noninductive overload, 177
"Normally," 52
Normally closed, def., 291
Normally open, def., 291
Nucleus, def., 291
Nucleus, 2, 17

Ohm, def., 291
Ohm, 8, 17
Ohmmeter, def., 291
Ohm's law, def., 291
Ohm's law, 12–16, 17, 18, 27–29
Oil safety switch, 211–12
Open circuit, def., 291
Out of phase, def., 291
Overload, def., 291
Overload: magnetic, 59–60, 61;
    thermal, 59
Overloads, 177–84; noninductive,
    177; fuses, 177; circuit breakers,
    177; line break, 178–79; pilot
    duty, 178, 181–83; internal com-
    pressor, 180–81; magnetic
    (heineman), 182–83; trouble-
    shooting, 183–84, 220–25
Overshoot, def., 291
Overshoot, 196

Packaged air-conditioning unit,
    def., 291
Packaged air-conditioning units,
    239–42; air-cooled, 239, 241,
    243–45; water-cooled, 239–40,
    246; control system for, 240–41;
    gas-electric, 242–43; rooftop
    units, 247
Parallel circuit, def., 291
Parallel circuit, 19, 25–29; applica-
    tions, 26; calculation for current,
    resistance, and voltage, 26–29
Peak voltage, def., 292

Peak voltage, 83
Permanent magnet, def., 292
Permanent split-capacitor motor,
    def., 292
Permanent split-capacitor motors
    (PSC), 136–38; operation, 136;
    troubleshooting, 136–38
Phase, def., 292
Phase, 84
Pictorial diagram, def., 292
Pictorial diagrams, 74–76
Pilot duty, def., 292
Pilot duty overload, 178, 181–83;
    troubleshooting, 222–23
Pneumatic control system, 273
Pole, def., 292
Pole, 51
Positive charge, def., 292
Potential coil, def., 292
Potential difference, def., 292
Potential difference, 8
Potential relay, def., 292
Potential relays, 152–53; operation,
    152–53; troubleshooting, 153
Power, 86
Power factor, def., 292
Power factor, 12
Pressure switch, def., 292
Pressure switches, 187, 202–7; high-
    pressure, 205; low-pressure, 205–
    6; notation and terms, 206–7;
    differential, 206; cut-in, 206; cut-
    out, 206; range, 206–7; trouble-
    shooting, 207
Proton, def., 292
Protons, 2, 17
Pump-down control system, def.,
    292
Pump-down control system, 270
Push-button switch, def., 292
Push-button switch, 55, 56, 163–
    64

Range, def., 292

Reactance, def., 292

Reactance, 84, 85

Relay, def., 292

Relays, 50, 51, 173–77; operation, 173–74; application, 174–75; construction, 175–76; troubleshooting, 176–77

Relief valve, def., 292

Remote bulb thermostat, 189, 190

Residential units, control systems, 262–63

Resistance, def., 292

Resistance, 11, 17; calculating for, series circuits, 23–25; calculating for, parallel circuits, 26–29; measurable, 39

Resistance heaters, 45, 46

Rotor, def., 292

Rotors, 123–24; squirrel cage, 124

Running (oil-filled) capacitor, def., 292

Running capacitor, 134

Safety controls, 74, 261–62

Safety device, def., 292

Safety devices, air conditioners, 58–60

Schematic diagram, def., 292

Schematic diagrams, air conditioners, 60–74; simple, 62–68; advanced, 68–74

Series circuit, def., 293

Series circuit, 19, 21–25; applications, 22–23; calculations for current, resistance, and voltage, 23–25

Series-parallel circuit, def., 293

Series-parallel circuit, 29–30

Shaded-pole motor, def., 293

Shaded-pole motor, 126–29; operation, 127–28; reversing, 128; troubleshooting, 128–29

Short circuit, def., 293

Short cycling, def., 293

Short cycling, 237

Signal light, def., 293

Signal lights, 49, 51

Sine wave, def., 293

Sine wave, 82

Single-phase motors, starting relays for, 150

Sleeve bearing, def., 293

Sleeve bearings, 155, 157, 158

Sliding armature, def., 293

Snap action of a thermostat, def., 191

Solenoid, def., 293

Solenoids, 45, 48, 49, 50

Solenoid valve, def., 293

Solenoid valves, 213–14

Split-phase motor, def., 293

Split-phase motors, 129–32; operation, 130–31; troubleshooting, 131–32

Squirrel cage rotor, def., 293

Squirrel cage rotor, 124

Staged system, def., 293

Staging thermostats, 199–202; operation and types, 201–2; heat pump, 202

Star transformer, def., 293

Star winding, def., 293

Starters, magnetic, 53, 54

Starting capacitor, def., 293

Starting capacitor, 133–34

Starting relay, def., 293

Starting relays, single-phase motors, 150

Static electricity, def., 293

Static electricity, 4–5

Stator, def., 293

Swinging armature, def., 293

Switch, def., 294
Switches, 53–58; disconnect, 55, 56, 107–10; manual, 54–55; push-button, 55, 56; thermostats, 55–57, 67; pressure, 57–58; enclosures, 108; fusible and nonfusible, 108–10; diagnosis, 275–80
System lag, def., 294
System lag, 196

Thermal overload, def., 294
Thermal overload, 59
Thermal relay, def., 294
Thermostat, def., 294
Thermostat controlling element, def., 294
Thermostats, 55–57, 67, 186, 187–202; applications, 187–88; controlling elements, 188–89; remote bulb, 189, 190; bimetal, 188, 190–92; snap action, 191; line voltage, 192, 194; low-voltage, 193–94, 195; installation, 198; staging, 199–202; trouble-shooting, 225–29
Three-phase motor, def., 294
Three-phase motors, 141–44; operation, 142; schematic diagrams, 143; troubleshooting, 144
Three-phase voltage system, 89
Throw, def., 294
Throw, 54
Time clock, 212–13
Time-delay relay, 212
Timer relay circuits, 70–71
Timers, 70–71
Torque, def., 294
Torque, 119
Transformer, def., 294

Transformers, 60, 61, 186, 207–10; operation, 207–8; sizing, 208; troubleshooting, 209–10, 229–30
208 volt-three-phase-60 hertz system, 91–92
230 volt-single phase-60 hertz systems, 87–89
230 volt-three phase-60 hertz systems, 90–91
277/480- volt system, 94–95

V-belt, def., 294
Volt, def., 294
Voltage, def., 294
Voltage, 8, 17; calculating for, series circuits, 23–25; line, 25; calculating for, parallel circuits, 26–29; peak, air conditioners, 83; effective, air conditioners, 83
Voltage-current systems, 83–84
Voltage drop, def., 294
Voltage drop, 101, 105–7
Voltage systems, 87–95
Voltmeter, def., 294

Water chiller controls, 264
Water chillers, def., 294
Water-cooled condenser interlocks, 267
Water-cooled packaged unit, def., 294
Watt, def., 294
Watthours (Wh), 12
Watts, 12, 17
Wire, sizing, 98–107
Wiring diagram, def., 294
Wiring diagrams, air conditioners, 44–80

Y transformer, def., 294